T0234782

Lecture Notes in Physics

Volume 963

The Lecture Notes in Physics

The series Lecture Notes in Physics (LNP), founded in 1969, reports new developments in physics research and teaching-quickly and informally, but with a high quality and the explicit aim to summarize and communicate current knowledge in an accessible way. Books published in this series are conceived as bridging material between advanced graduate textbooks and the forefront of research and to serve three purposes:

- to be a compact and modern up-to-date source of reference on a well-defined topic
- to serve as an accessible introduction to the field to postgraduate students and nonspecialist researchers from related areas
- to be a source of advanced teaching material for specialized seminars, courses and schools

Both monographs and multi-author volumes will be considered for publication. Edited volumes should however consist of a very limited number of contributions only. Proceedings will not be considered for LNP.

Volumes published in LNP are disseminated both in print and in electronic formats, the electronic archive being available at springerlink.com. The series content is indexed, abstracted and referenced by many abstracting and information services, bibliographic networks, subscription agencies, library networks, and consortia.

Proposals should be sent to a member of the Editorial Board, or directly to the managing editor at Springer:

Dr Lisa Scalone
Springer Nature
Physics Editorial Department
Tiergartenstrasse 17
69121 Heidelberg, Germany
lisa.scalone@springernature.com

More information about this series at http://www.springer.com/series/5304

Valery Zagrebaev

Heavy Ion Reactions at Low Energies

Andrey Denikin • Alexander Karpov • Neil Rowley

Editors

 Springer

Author
Valery Zagrebaev
Flerov Laboratory of Nuclear Reactions
Joint Institute for Nuclear Research
Dubna, Russia

Editors
Andrey Denikin
Nuclear Physics Department
Dubna State University
Dubna, Russia

Alexander Karpov
Flerov Lab of Nuclear Reactions
Joint Institute for Nuclear Research
Dubna, Russia

Neil Rowley
Institut de Physique Nucléaire
UMR 8608, CNRS-IN2P3 and Université de
Orsay CEDEX, France

Translated by
Andrey Denikin
Dubna, Russia

Alexander Karpov
Dubna, Russia

ISSN 0075-8450 ISSN 1616-6361 (electronic)
Lecture Notes in Physics
ISBN 978-3-030-27216-6 ISBN 978-3-030-27217-3 (eBook)
https://doi.org/10.1007/978-3-030-27217-3

Translation from the Russian language edition: Ядерные реакции с тяжелыми ионами by
Загребаев В.И., © V.I. Zagrebaev 2016, ISBN 978-5-9530-0435-0. Published by Joint Institute for
Nuclear Research. All Rights Reserved.

This Springer imprint is published by the registered company Springer Nature Switzerland AG.
The registered company address is: Gewerbestrasse 11, 6330 Cham, Switzerland

Foreword

More than a 100 years of advances in the field of nuclear physics have allowed scientists to frame a general idea of the structure of atomic nuclei, to determine the main characteristics of nuclear interactions, to answer questions on the origin of the chemical elements in the Universe, and to learn how to produce new elements and predict their properties. Modern nuclear physics has provided a wealth of practical experience and many tools that enable us to solve a wide range of applied problems that humanity currently faces. All this makes it even more intriguing that after such long-term research in this field, many blank pages remain to be filled.

Over the past few decades, new experimental methods have been developed that allow us to reveal areas of nuclear physics previously inaccessible to investigation. The boundaries of the known parts of the nuclear map have been considerably extended. We can now obtain and investigate systems of nucleons with very exotic properties, in particular neutron-rich nuclei located at the drip line, such as ^{7}H, ^{8,10}He, ^{22}C, and ^{28}O. In addition, an increase in experimental sensitivity has made it possible to study, for example, nuclear reactions occurring at deep sub-barrier energies, and a breakthrough in the studies of the heaviest nuclei has also been made. In the last 15 years, six new superheavy elements have been synthesized, and convincing evidence for an "island of stability" in this region of mass—predicted over half a century ago—has finally been established.

The rapid developments in nuclear and elementary particle physics over recent years require the information accumulated over a 30-year period to be systematized and presented comprehensively. Unfortunately, the scope of recently published Russian academic literature on this subject is somewhat lacking. The author of this tutorial—a scientist with an excellent international reputation—has managed to partially fill this gap. This book outlines the main experimental facts on nuclear reactions involving heavy ions at low energies. It focuses on discussions of nuclear physics processes that are a subject of active, modern research, and it gives pictorial explanations of these phenomena in the framework of up-to-date theoretical concepts.

The book is intended for physics students who have completed a standard course of quantum mechanics and have a basic idea of nuclear physics processes. It is designed to become a kind of lifeboat that at the end of the course will allow students to navigate the modern scientific literature and to understand the goals and objectives of current, on-going research.

Dubna, Russia   Andrey Denikin
Dubna, Russia   Alexander Karpov
2015

Contents

Chapter 1
Introduction

To begin with, let us briefly recall the basic properties of atomic nuclei, such as binding energy, size, and shape (deformation). The mass of the nucleus is always less than the sum of the masses of its constituent nucleons. The difference between these values is called *the nuclear binding energy*:

$$E_{\text{bind}}(Z, A) = Z \cdot m_p c^2 + N \cdot m_n c^2 - M(Z, A)c^2, \qquad (1.1)$$

where Z and N are the numbers of protons and neutrons in the nucleus, m_p and m_n are their masses, $A = Z + N$ is total number of nucleons. In the low-energy nuclear reactions that we consider here, the number of protons and neutrons remains unchanged. The binding energies of the nuclei in the initial and final states often play a key role in determining the probability and behavior of a particular reaction. For example, the law of energy conservation in the binary reaction $a + A \rightarrow b + B$ (in which a nucleus a hits a fixed target A) is written as:

$$M(a)c^2 + M(A)c^2 + E_{\text{kin}}(a) = M(b)c^2 + M(B)c^2 + E_{\text{kin}}(b) + E_{\text{kin}}(B),$$

and it is the nuclear binding energy that determines the kinetic energy of the particles formed in the reaction (here it is assumed that the nuclei b and B are formed in the ground state, otherwise their excitation energies should be added to the right side):

$$E_{\text{kin}}(b) + E_{\text{kin}}(B) = E_{\text{kin}}(a) + [E_{\text{bind}}(b) + E_{\text{bind}}(B) - E_{\text{bind}}(a) - E_{\text{bind}}(A)],$$

and the quantity $Q = E_{\text{bind}}(b) + E_{\text{bind}}(B) - E_{\text{bind}}(a) - E_{\text{bind}}(A)$ is called the reaction Q-value. Since kinetic energy cannot be negative, when $Q < 0$, this reaction is possible only for $E_{\text{kin}}(a) > |Q|$ (endothermic reaction). Reactions with $Q > 0$ are called exothermic. An example of exothermic reactions (those with

© Springer Nature Switzerland AG 2019
V. Zagrebaev, *Heavy Ion Reactions at Low Energies*, Lecture Notes in Physics 963,
https://doi.org/10.1007/978-3-030-27217-3_1

energy release) is fusion reactions of light nuclei

$$^2\text{H} + {}^3\text{H} \rightarrow {}^4\text{He} + n + 14\,\text{MeV}$$

and fission reactions of heavy nuclei

$$n + {}^{235}\text{U} \rightarrow {}^{102}\text{Mo} + {}^{132}\text{Mo} + 2n + 190\,\text{MeV},$$

whereas fission reactions of light nuclei and fusion reactions of heavy nuclei are endothermic. That is, to make them occur an additional kinetic energy of the colliding particles is required:

$$^{16}\text{O} + {}^{208}\text{Pb} \rightarrow {}^{12}\text{C} + {}^4\text{He} + {}^{208}\text{Pb} - 7\,\text{MeV},$$

$$^{48}\text{Ca} + {}^{248}\text{Cm} \rightarrow {}^{293}116 + 3n - 160\,\text{MeV}.$$

This can be explained by the characteristic dependence of the specific binding energy $\varepsilon_{\text{bind}}(A) = E_{\text{bind}}(A)/A$ on the mass number. This dependence is illustrated in Fig. 1.1.

At first, the specific binding energy increases sharply with increasing nucleon number; it reaches its maximum at $A \sim 60$ and then gradually decreases with increasing A. This means that in the process of fission of heavy nuclei, the fragments that are formed are more strongly bound, because the sum of their masses is less than that of the initial nucleus. Similarly, in the fusion of light nuclei, a more strongly bound nucleus is produced, its mass being less than the sum of the masses of the initial nuclei. This is the reason why these reactions are accompanied by energy release. This behavior of the binding energy per nucleon seen in Fig. 1.1 can be explained by the short-range nature of the nuclear attraction (surface nucleons, that are more numerous in light nuclei, have fewer neighbors and thus make a smaller contribution to the binding energy) and by an increase in the repulsive Coulomb force with increasing charge of the nucleus. These effects are reflected in Weizsacker's formula [70]

$$E_{\text{bind}}(Z, A) \approx c_{\text{tot}}A - c_{\text{surf}}A^{2/3} - \frac{3}{5}\frac{Z^2 e^2}{R_C} - \frac{1}{2}c_{\text{sym}}\frac{(N - Z)^2}{A} \tag{1.2}$$

with the coefficients $c_{\text{tot}} \approx 15.6\,\text{MeV}$, $c_{\text{surf}} \approx 17.2\,\text{MeV}$, and $c_{\text{sym}} \approx 46.6\,\text{MeV}$, which provides a sufficiently qualitative approximation of the binding energy (see Fig. 1.1).

Due to the short-range of the nuclear forces each nucleon interacts strongly only with its immediate neighbors, and the first term in Eq. (1.2) reflects the binding energy of all the nucleons due to their nearest-neighbor interactions. The second term takes into account the fact that the surface nucleons (their number being proportional to the surface area of the nucleus $S = 4\pi R^2 \sim A^{2/3}$) have fewer neighbors and, therefore, are less bound. That is, it is necessary to slightly reduce

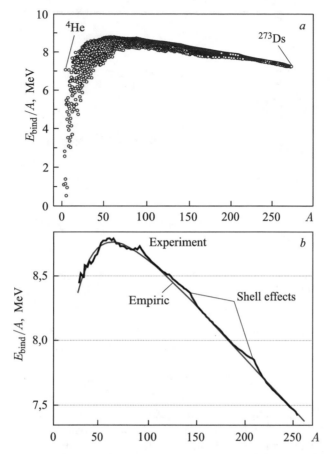

Fig. 1.1 (a) Specific binding energies of nuclei, experimental values. (b) Specific binding energies of nuclei located along the drip line, experimental values, and values calculated by Weizsacker's formula (smooth curve)

the total binding energy (hence the minus sign). The third term takes into account the Coulomb repulsion of the protons, which further reduces the binding energy. Shell effects (disregarded in this formula) make a notable contribution to the total binding energy of the nucleus (more than 10 MeV for ^{208}Pb) and play an important role in the dynamics of nuclear reactions at low energies (see below).

Determining the size of the nucleus is quite a challenge, although the well-known Rutherford experiment suggested that the radius of the gold nucleus did not exceed 10^{-12} cm. The most direct measurement of the charge distribution inside the nucleus was made in experiments on the scattering of 153 MeV electrons, also by gold nuclei [29]. The resulting (albeit poorly manifested) diffraction picture made it possible to conclude a fairly uniform charge distribution inside the gold nucleus with a radius of about 6.5 fm and diffuseness of 0.5 fm. Further experiments (including ones using laser spectroscopy) allowed the determination of the charge radii of most nuclei near

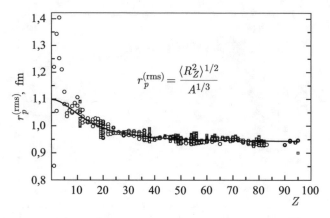

Fig. 1.2 Experimental values of the mean charge radius divided by $A^{1/3}$. The solid curve corresponds to formula (1.3)

the drip line. Figure 1.2 shows the experimental values of the mean nuclear radius divided by $A^{1/3}$. It can be seen that the mean-square charge radii of medium and heavy nuclei are well approximated by the formula $\langle R_p^2 \rangle^{1/2} = r_p^{(rms)} A^{1/3}$, where $r_p^{(rms)} \approx 1$ fm. Taking into account an increase in $r_p^{(rms)}$ with decreasing Z, one can propose the following approximate formula for this value:

$$r_p^{(rms)} = 0.94 + \frac{32}{Z^2 + 200} \text{ fm.} \tag{1.3}$$

This is obtained by simple fitting of the experimental data (see the solid curve in Fig. 1.2) and is applicable for nuclei heavier than carbon.

Assuming a uniform density of matter in the spherical nucleus up to a radius R, that is, $\rho(r \leqslant R) = \rho_0 = A \big/ \frac{4}{3}\pi R^3$ and $\rho(r > R) = 0$, the mean radius is less than R; in particular, $\langle R^2 \rangle^{1/2} = \sqrt{3/5}R$. The diffuseness a of the nuclear surface provides a smoothing of the decrease in density around $r = R$. With a finite diffuseness, a better approximation for the charge radius $R_p = r_p^0 A^{1/3}$ is given by

$$r_p^0 = r_p^{(rms)} \sqrt{\frac{5}{3}\left(1 - \frac{7}{5}\pi\xi + O(\xi^3)\right)}, \tag{1.4}$$

where $\xi = a/r_p^{(rms)} A^{1/3}$.

Direct measurements of the nuclear matter radius are not possible and the most informative experiments are those on diffraction scattering of protons and neutrons. The results may then be analyzed, for example, in the framework of the optical model (see below). An example of diffraction scattering of protons by calcium nuclei is shown in Fig. 1.3a. Assuming that the diffraction is produced by a semi-transparent sphere of radius R_A, and knowing the de Broglie wavelength of the proton $\lambdabar = \hbar/\sqrt{2m_p E_p}$, we can roughly estimate the value of R_A from the position

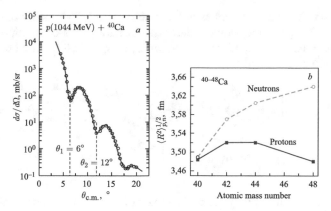

Fig. 1.3 (**a**) Elastic scattering of protons with an energy of 1044 MeV by calcium nuclei [3]. (**b**) Mean nuclear radii of calcium isotopes obtained from the analysis of diffraction scattering of protons [2]

of the diffraction minima using the formula $2R_A \sin\theta_n = n\lambda$ ($\lambda = 2\pi\,\hbar$), where n is the number of the minimum. At high energies, it is necessary to take into account the relativistic correction to the proton mass. From Fig. 1.3a we obtain an approximate value of 3.4 fm for the radius of ^{40}Ca. A more accurate estimate can be made by accurately analyzing these data within the framework of a quantum mechanical model of elastic neutron scattering (see below). The analysis of a large amount of experimental data allows us to conclude that the distribution of neutrons in the nucleus is also fairly uniform. For light and medium nuclei located near the drip line, the distribution of neutrons and protons inside the nucleus is actually the same, that is, $r_n^0 \approx r_p^0$. In heavy nuclei, the number of neutrons significantly exceeds the number of protons and $R_N > R_Z$. This is already noticeable for neutron-rich calcium isotopes (see Fig. 1.3b). If we assume that the neutron and proton mean densities are approximately identical in heavy nuclei $\rho_0^p \approx \rho_0^n$, i.e., $N/R_N^3 \approx Z/R_Z^3$, then $R_N/R_Z \approx (N/Z)^{1/3}$.

The shape of the majority of nuclei in the ground state is close to spherical (minimal surface for a given volume). However, for many nuclei a deviation (deformation) from the spherical shape can be observed in the ground state, whereas dynamical deformations of the nuclear surface (vibrations) are possible for any nucleus. The deviation from the spherical shape is defined by the expression

$$
R(\vec{\beta}; \theta, \varphi) = \tilde{R}_0 \left(1 + \sum_{\lambda \geq 2} \sum_{\mu=-\lambda}^{\lambda} \beta_{\lambda\mu} Y_{\lambda\mu}(\theta, \varphi) \right)
$$

$$
= \tilde{R}_0 \left(1 + \sum_{\lambda \geq 2} \beta_\lambda \sqrt{\frac{2\lambda + 1}{4\pi}} P_\lambda(\cos\theta) \right), \tag{1.5}
$$

where the second equation applies in the case of axially symmetric deformations. Here $\vec{\beta} \equiv \{\beta_\lambda\}$ are dimensionless deformation parameters of multipolarity $\lambda = 2, 3, \ldots$, the functions $Y_{\lambda\mu}(\Omega)$ are the orthonormal spherical harmonics, P_λ are the Legendre polynomials and

$$\tilde{R}_0 = R_0 \left[1 + \frac{3}{4\pi} \sum_\lambda \beta_\lambda^2 + \frac{1}{4\pi} \sum_{\lambda,\lambda',\lambda''} \sqrt{\frac{(2\lambda'+1)(2\lambda''+1)}{4\pi(2\lambda+1)}} \times \right.$$
$$\left. \times (\lambda'0\lambda''0|\lambda0)^2 \beta_\lambda \beta_{\lambda'} \beta_{\lambda''} \right]^{-1/3}, \tag{1.6}$$

where R_0 is the radius of a sphere of the same volume as the deformed nucleus, and $(\lambda'0\lambda''0|\lambda0)$ are Clebsch–Gordon coefficients. Examples of nuclei with quadrupole $(\lambda = 2)$, octupole $(\lambda = 3)$, and hexadecupole $(\lambda = 4)$ deformations are illustrated in Fig. 1.4. The presence of a static quadrupole deformation β_2^{gs} in the nucleus means that it has a quadrupole moment $Q_0 = \frac{3}{\sqrt{5\pi}} Z R_0^2 \beta_2^{gs}$, and this can be measured experimentally.

In the case of dynamical deformations, for example, the excited state $3^-(2.6\,\text{MeV})$ of the ^{208}Pb nucleus, one can define the mean value of the deformation

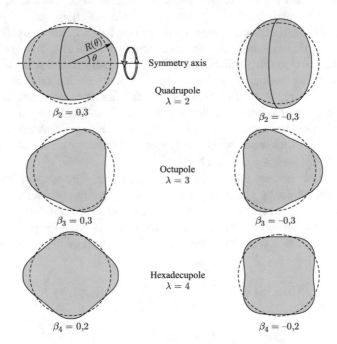

Fig. 1.4 Shape of nuclei with quadrupole, octupole, and hexadecupole deformations of the surface (on the left-hand side positive values of the deformation parameters, on the right, negative values)

of the corresponding oscillations $\langle \beta_\lambda^0 \rangle = \frac{4\pi}{3ZR_0^\lambda} \left[\frac{B(E\lambda)}{e^2} \right]^{1/2}$, which can be found from the experimental value of the probability $B(E\lambda)$ of electromagnetic transitions from this excited state to the ground state. The rigidity of the nuclear surface in relation to oscillations of multipolarity λ is given by the expression $C_\lambda = (2\lambda + 1)\frac{\varepsilon_\lambda}{2\langle \beta_\lambda^0 \rangle^2}$, where $\varepsilon_\lambda = \hbar \omega_\lambda$ is the associated energy.

Chapter 2
Nuclear Interactions and Classes of Nuclear Reaction

2.1 Nucleon–Nucleon and Nucleon–Nucleus Interactions, Nuclear Mean Field

Despite the long history of nuclear physics, we are still unable to write down, in simple terms, the *exact* formula for the two-nucleon interaction. The reason lies in the complex nature of nuclear forces and in the complex structure of the nucleons themselves, which, at the low energies of interest to us, behave as indestructible, elementary particles. The discussion of this problem is beyond the scope of this course, therefore we will confine ourselves to stating the currently known properties of the nucleon–nucleon interaction.

From the well-established properties of the deuteron (^{2}H: spin $S_d = 1$, binding energy $E^d_{\text{bind}} = 2.2\,\text{MeV}$, radius $R_d \sim 4\,\text{fm}$, quadrupole moment $Q_d \approx 0.3\,\text{fm}^2$), we can make the following definite conclusions on the nature of the nucleon–nucleon force: (1) the short-range of its action, (2) its large magnitude (strength), (3) its dependence on the spin (shown by the absence of a bound $n-p$ state with zero spin), and (4) its non-centrality (shown by the fact that the deuteron has a small, nonzero quadrupole deformation Q_d). Scattering neutrons from protons gives additional information about the $n - p$ interaction. In particular, an unexpectedly large cross section for back-angle scattering has been observed (see Fig. 2.1). It indicates the exchange nature of nucleon–nucleon forces, that is, neutrons flying in the center-of-mass system in the opposite direction are formed from protons as a result of the exchange of charged interaction carriers. If the carriers are π-mesons, the range of nucleon–nucleon forces is determined from the following simple considerations. After emitting the π-meson, it must be absorbed within the time $\tau \sim \hbar/m_\pi c^2$ in order not to violate the law of energy conservation. During this time, it cannot move a distance greater than $r_{NN} = c\tau = \hbar/m_\pi c \approx 1.43\,\text{fm}$. This is the characteristic range of the action of internucleon forces.

In specific calculations (depending on what these calculations are aimed at) one uses either nucleon–nucleon forces calculated on the basis of the meson

© Springer Nature Switzerland AG 2019
V. Zagrebaev, *Heavy Ion Reactions at Low Energies*, Lecture Notes in Physics 963,
https://doi.org/10.1007/978-3-030-27217-3_2

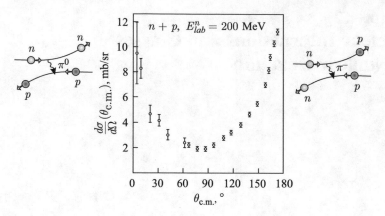

Fig. 2.1 Angular distribution of elastic neutron scattering by protons [38]

theory or phenomenological forces with parameters selected from a comparison with experiment. One of the phenomenological NN–potentials is, for example, the Hamada–Johnston potential

$$v_{NN}(r) = V_C(r) + V_T(r)S_{12} + V_{LS}(r)(\mathbf{L} \cdot \mathbf{S}) + V_{LL}(r)L_{12}, \tag{2.1}$$

which is the sum of the central (V_C), tensor (V_T), and spin-orbit forces of first (V_{LS}) and second (V_{LL}) order. In the above expression

$$S_{12} = \frac{3}{r^2}(\vec{\sigma}_1 \cdot \mathbf{r})(\vec{\sigma}_2 \cdot \mathbf{r}) - (\vec{\sigma}_1 \cdot \vec{\sigma}_2),$$

where $\vec{\sigma}_{1,2}$ are the spin operators of nucleons (Pauli matrices), $\mathbf{S} = \frac{1}{2}(\vec{\sigma}_1 + \vec{\sigma}_2)$ is the total spin of the two nucleons and $L_{12} = [\delta_{LJ} + (\vec{\sigma}_1 \cdot \vec{\sigma}_2)]\mathbf{L}^2 - (\mathbf{L} \cdot \mathbf{S})^2$. The radial dependence of the different components is taken as

$$V_C = v_0(\vec{\tau}_1 \cdot \vec{\tau}_2)(\vec{\sigma}_1 \cdot \vec{\sigma}_2)Y(x)[1 + a_C Y(x) + b_C Y^2(x)],$$

$$V_T = v_0(\vec{\tau}_1 \cdot \vec{\tau}_2)Z(x)[1 + a_T Y(x) + b_T Y^2(x)],$$

$$V_{LS} = g_{LS}v_0 Y^2(x)[1 + b_{LS}Y(x)],$$

$$V_{LL} = g_{LL}v_0 Z(x)/x^2[1 + a_{LL}Y(x) + b_{LL}Y^2(x)],$$

$$Y(x) = \exp(-x)/x, \quad Z(x) = (1 + 3/x + 3/x^2)Y(x),$$

$$x = r/(\hbar/m_\pi c) \approx r/1.43,$$

where $v_0 = 3.65$ MeV and $\vec{\tau}_{1,2}$ are the isospin operators. At small distances ($r < 0.48$ fm) the potential (2.1) is assumed to be equal to $+\infty$ (repulsive core). The remaining parameters (and other types of NN-potentials) can be found in the book

[22]. When describing the properties and interactions of heavy nuclei, it is usually assumed that the nucleon–nucleon forces in the nuclear medium can be different from their *vacuum* values. In this case, the so-called zero-range Skyrme forces are often used (see, for example, review [11]), which greatly facilitates the calculation of various matrix elements.

When describing the nucleon–nucleus interaction, the concept of the mean nuclear field has become widely accepted. It is based both on the experimental data on the elastic nucleon scattering by nuclei, and on the manifestation of the shell structure and single-particle bound states of atomic nuclei. If we calculate the density of nucleons in the nucleus by the formula $\rho_0 = A \big/ \frac{4}{3} \pi R_A^3$, using the expressions $R_A = r_0 A^{1/3}$ and $r_0 \sim 1.2\,\text{fm}$ for the radius of the nucleus, we will come to the conclusion that atomic nuclei are quite *hollow* inside: $\rho_0 \approx 0.2\,\text{fm}^{-3}$, whereas the volume of one nucleon does not exceed $1\,\text{fm}^3$. Consequently, as shown by the experimental data, it is highly probable that the nucleons passing through the nucleus deviate due to the attractive nuclear forces (refraction) and remain in the elastic channel (Fig. 2.2a). If we determine the probability of nucleon escape from the elastic channel (as a result of a collision with another nucleon or some inelastic excitation of the nucleus) by the expression

$$P_{abs} = 1 - \exp\left(-\int\limits_{tr} \frac{ds}{\lambda_{free}}\right),$$

in which the integral is calculated along the trajectory and λ_{free} denotes the nucleon free path in the nucleus, we can obtain the estimate of $\lambda_{free} \sim R_A$ from the experimental data analysis.

Such *semi-transparency* of the nucleus allows the use of the so-called optical model (OM) to describe elastic nucleon scattering by nuclei, in which the nucleon–nucleus interaction is determined by the optical potential (OP) $V(r) + iW(r)$. The real part of this expression determines the refraction (elastic scattering) of nucleons

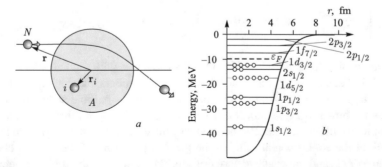

Fig. 2.2 (a) Elastic nucleon scattering by the mean nuclear field. (b) Single-particle neutron levels in the nucleus of ^{40}Ca calculated for the mean field with the depth $V_0 = -50\,\text{MeV}$, radius $r_0 = 1.25\,\text{fm}$ and diffuseness $a = 0.7\,\text{fm}$

by the mean nuclear field, and the imaginary part $W \sim \hbar v/2\lambda_{free}$ describes the loss of particles from the elastic channel (absorption). The OP parameters are selected from the analysis of the experimental data on elastic scattering within the framework of OM (see below). However, it is possible to obtain a reasonable estimate of $V(r)$ if we represent the nucleon–nucleus interaction as a sum of nucleon–nucleon interactions averaged over the nucleon densities given by their single-particle wavefunctions (see Fig. 2.2a):

$$V(r) = \sum_{i=1}^{A} \int v_{Ni}(r - r_i)|\varphi_i(r_i)|^2 d^3 r_i. \tag{2.2}$$

Approximating the short-range nucleon–nucleon interaction by delta forces $v_{Ni}(r - r_i) = -u_0\delta(r - r_i)$, for $V(r)$ we obtain a simple expression

$$V(r) = -u_0 \sum_{i=1}^{A} |\varphi_i(r)|^2 = -u_0\rho(r) \approx \frac{-V_0}{1 + \exp\left(\frac{r - R_A}{A}\right)},$$

where $\rho(r)$ is the nucleon density, which, as noted above, is essentially constant inside the nucleus and rapidly decreases to zero in the boundary layer $a \approx 0.6\,\text{fm}$. The depth of the nucleon–nucleus mean field V_0 in this case is about 50 MeV.

The concept of the nuclear mean field is also reflected in the independent-particle model that describes the bound states of nucleons within the atomic nucleus. The main assumption of this model is that the nucleons move independently in the mean field generated by the nucleon–nucleon interactions averaged over their relative motion. In this case, the total wavefunction of the nucleus can be written as an antisymmetrized product of single-particle wavefunctions $\Psi_A(r_1, r_2, \ldots, r_A) = \left\{\prod_{i=1}^{A} \varphi_i(r_i)\right\}_{\pm}$, which are determined by solving the single-particle Schrödinger equation

$$\left[-\frac{\hbar^2}{2m}\nabla^2 + V(r) + V_{SO}(r)(\vec{\ell} \cdot \vec{s})\right]\varphi_{nlj\mu}(r) = \varepsilon_{nlj}\varphi_{nlj\mu}(r). \tag{2.3}$$

The spin–orbit interaction $V_{SO}(r)(\vec{\ell} \cdot \vec{s})$ splits the levels with total angular momenta $j = \ell \pm 1/2$ and correctly reproduces the nucleon numbers that correspond to filled shells (magic numbers). An example of neutron single-particle states in the ^{40}Ca nucleus is shown in Fig. 2.2b). It can be seen from the figure that the shell $1d_{3/2}$ is completely filled and the 21st neutron can only be in the state $1f_{7/2}$. Indeed, if we look at the experimental value of the ground state spin of the ^{41}Ca nucleus, it is equal to 7/2, and the negative parity of this state indicates the orbital momentum $\ell = 3$.

2.2   Nucleus–Nucleus Interaction: Folding and Phenomenological Potentials

The potential energy of the nucleus–nucleus interaction, as with any physical system, is the key characteristic that determines the properties of the system and its evolution with time. At low excitation energies in a heavy nucleus (consisting of tens or hundreds of nucleons), only a few collective degrees of freedom play a role. To describe the system properly, the correct choice of these degrees of freedom is essential. The distance between the centers of the nuclei, their mutual orientation, and the dynamical deformations of their surfaces are the principal coordinates to consider in low-energy heavy-nuclear collisions. The nature of the nucleus–nucleus interaction also depends on the relative velocity of the nuclei. When this exceeds the velocity of a nucleon within the nucleus (i. e., the Fermi velocity), the nucleus–nucleus potential must contain a repulsive component at short distances that prevents the overlapping of the initial *frozen* densities that would be accompanied by the formation of a region with double density (region marked in dark in Fig. 2.3). Such collision conditions are called diabatic. In slower collisions ($v_{rel} \ll v_{Fermi}$) the nucleons have enough time to reach the equilibrium distribution in a volume with constant density (adiabatic conditions), and the nucleus–nucleus potential looks very different (Fig. 2.3). These conditions are obviously satisfied at the collision energies close to the height of the Coulomb barrier, when the relative velocity of the nuclei in the barrier region (i.e., before they make contact) is close to zero. The adiabatic potential energy and methods for its calculation are discussed in Chap. 5 below. For nuclei that have separated, the diabatic and adiabatic potentials naturally coincide.

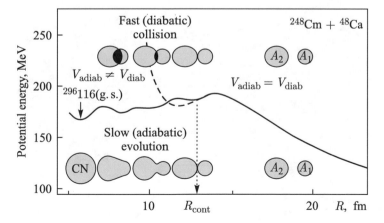

Fig. 2.3 Potential energy in diabatic (dashed curve) and adiabatic (solid curve) nuclear collisions of ^{48}Ca and ^{248}Cm

2.2.1 Folding Potentials

The most consistent approach to the calculation of the diabatic nucleus–nucleus interaction is the so-called folding procedure, where we make a simple summation and average of the effective nucleon–nucleon interaction over the nuclear densities (see, for example, [58]). In this case, the effects due to the curvature of the nuclear surfaces are automatically taken into account. Using the same approach, the internuclear interaction can also be calculated relatively simply for an arbitrary orientation of statically deformed nuclei (see Fig. 2.4). In this case, the interaction potential can be written as

$$V_{12}(R; \vec{\beta}_1, \Omega_1, \vec{\beta}_2, \Omega_2) = \int_{V_1} \rho_1(\mathbf{r}_1) \int_{V_2} \rho_2(\mathbf{r}_2) v_{NN}(\mathbf{r}_{12}) \, d^3\mathbf{r}_1 \, d^3\mathbf{r}_2, \qquad (2.4)$$

where $v_{NN}(\mathbf{r}_{12} = \mathbf{R} + \mathbf{r}_2 - \mathbf{r}_1)$ is the effective nucleon–nucleon potential; $\rho_i(\mathbf{r}_i)$ is the density of the nuclear matter distribution in the i-th nucleus, $\Omega_i = \{\theta_i, \phi_i\}$ are the spherical angles defining the nuclear orientation, and $\vec{\beta}_i \equiv \{\beta_\lambda^{(i)}\}$ are dimensionless parameters relating to the deformation of the i-th nucleus (see Introduction). The nuclear matter density $\rho_i(\mathbf{r})$ is usually taken as a Fermi function with diffuseness a_i:

$$\rho_i(\mathbf{r}) = \frac{\rho_0}{1 + \exp\left(\frac{r - R_i(\beta, \Omega_{\mathbf{r}})}{a_i}\right)}, \qquad (2.5)$$

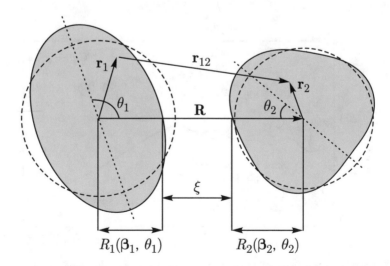

Fig. 2.4 Schematic view of two deformed nuclei rotated in the reaction plane. The angles θ_1 and θ_2 are the orientations of the symmetry axes of the two nuclei

where $R_i(\beta, \Omega_\mathbf{r})$ is the radius of a point on the nuclear surface (with $\Omega_\mathbf{r}$ the spherical coordinates of the vector $\mathbf{r}$). The value of ρ_0 is determined from the normalization condition $\int \rho_i d^3\mathbf{r} = A_i$.

The effective nucleon–nucleon potential consists of a short-range nuclear part and a long-range Coulomb part: $v_{NN} = v_{NN}^N + v_{NN}^C$, where the latter, $v_{NN}^C(r) = e^2/r$, applies only between the protons. For the nuclear part of the effective nucleon–nucleon interaction, both finite and zero-range approximations have been used. The density-dependent, zero-range effective interaction greatly simplifies the calculation of the six-dimensional integral in Eq. (2.4), and with an appropriate parameterization of the nuclear density provides correct heights and positions of the experimental Coulomb barriers for nucleus–nucleus reactions.

Among the finite-range potentials the most frequently used is the M3Y interaction [58]. With an appropriate choice of parameters, the folding M3Y potential describes the Coulomb barriers of nuclei quite well and may also be successfully used to describe elastic scattering. However, in the region of nuclear overlap, where repulsion due to the Pauli principle should occur and prevent the occurrence of a region with double nuclear density (incompressibility of nuclear matter), the M3Y-interaction generates too strong attraction. This, in particular, contradicts experiments on the deep-inelastic scattering of heavy nuclei. A solution to this problem is to introduce an additional, phenomenological, zero-range, repulsive term into the effective nucleon–nucleon interaction, bringing an additional contribution $V_0^{rep} \int_{V_1} \rho_1(\mathbf{r}_1, \tilde{a}_1, r_{01})\rho_2(\mathbf{R} - \mathbf{r}_1, \tilde{a}_2, r_{02})d^3\mathbf{r}_1$ to the expression (1.4). In this term, the diffuseness parameters in the nuclear densities are recommended to be slightly lower, for example, $\tilde{a} = 0.65\,a$, so that the repulsive part does not exert a strong influence on the Coulomb barrier height. The value of the repulsive potential $V_0^{rep} \approx 480$ MeV $\cdot$ fm^3 is chosen to reproduce the incompressibility coefficient.

The density- and energy-dependent effective nucleon–nucleon M3Y-potential includes a direct and an exchange term:

$$v_{M3Y}^N(\mathbf{r}) = [v_{M3Y}^{dir}(\mathbf{r}) + v_{M3Y}^{ex}(\mathbf{r})]F(\rho)\left(1 - k\frac{E}{A}\right), \tag{2.6}$$

where $k = 0.003$ MeV^{-1} and E is the collision energy. For the direct part of the interaction it is recommended to use the expression

$$v_{M3Y}^{dir}(r) = 11061.625\frac{\exp(-4r)}{4r} - 2537.5\frac{\exp(-2.5r)}{2.5r}, \tag{2.7}$$

whereas for the exchange term a zero-range force is preferred $v_{M3Y}^{ex} = -592\delta(\mathbf{r})$. The function

$$F(\rho) = C[1 + \alpha \exp(-\beta\rho) - \gamma\rho], \tag{2.8}$$

with the parameters $C = 0.2658$, $\alpha = 3.8033$, $\beta = 1.4099\,\text{fm}^3$, $\gamma = 4.0\,\text{fm}^3$, and $\rho(\mathbf{r}) = \rho_1(\mathbf{r}) + \rho_2(\mathbf{r})$ determines the density dependence of the nucleon–nucleon interaction.

To reduce the number of required parameters, one can use proton and neutron densities with identical radii $r_p^0 = r_n^0$ and diffuseness $a_p = a_n = a$. The global parameterization of the charge radius r_p^0 can be obtained by approximating the available experimental data (see Fig. 1.2 and formulas (1.3) and (1.4)). The value of the second parameter of the model, the diffuseness a, is chosen to obtain the best correspondence of the derived fusion barriers with their experimental values or with the empirical *Bass barriers* [7]. In particular, the expression

$$a(Z, A) = 0.45 - 6(Z^2/A)^{-3}, \qquad (2.9)$$

can be recommended for the diffuseness parameter of nuclear densities, when calculating the folding potentials for $A_{1,2} \geqslant 16$ and $Z_{1,2} \geqslant 8$.

2.2.2 Woods–Saxon Potential

In most cases, the potential energy for the interaction between two separated nuclei can be parameterized by some simple function, for example, the Woods–Saxon potential or the proximity potential [13]. In the case of insignificant deformations, the shape of the axially symmetric nucleus is usually determined by the formula (1.5), and the potential energy between two deformed nuclei, shown in Fig. 2.4, can be written as a sum of Coulomb and nuclear terms:

$$V_{12}(R; \vec{\beta}_1, \theta_1, \vec{\beta}_2, \theta_2) = V_C(R; \vec{\beta}_1, \theta_1, \vec{\beta}_2, \theta_2) + V_N(R; \vec{\beta}_1, \theta_1, \vec{\beta}_2, \theta_2). \qquad (2.10)$$

Neglecting the multipole–multipole interaction, to second-order accuracy, the Coulomb interaction of two deformed nuclei can be written

$$V_C = Z_1 Z_2 e^2 \left[F^{(0)}(R) + \sum_{i=1}^{2} \sum_{\lambda \geqslant 2} F_{i\lambda}^{(1)}(R) \beta_{i\lambda} Y_{\lambda 0}(\theta_i) \right] +$$

$$+ Z_1 Z_2 e^2 \sum_{i=1}^{2} \sum_{\lambda'} \sum_{\lambda''} \sum_{\lambda=|\lambda'-\lambda''|}^{\lambda=\lambda'+\lambda''} F_{i\lambda}^{(2)}(R) \sum_{\mu} \int Y_{\lambda'\mu}^* Y_{\lambda''-\mu}^* Y_{\lambda 0} d\Omega \times$$

$$\times \beta_{i\lambda'} \beta_{i\lambda''} Y_{\lambda 0}(\theta_i) + \ldots \qquad (2.11)$$

Here $F_\lambda^{(n)}(R)$ are the interaction form factors. When $R > R_1 + R_2$, we have $F^{(0)} = \frac{1}{R}$, $F_{i\lambda}^{(1)} = \frac{3}{2\lambda+1} \frac{R_i^\lambda}{R^{\lambda+1}}$, $F_{i\lambda=2}^{(2)} = \frac{6}{5} \frac{R_i^2}{R^3}$, $F_{i\lambda=4}^{(2)} = \frac{R_i^4}{R^5}$. When the values of R are lower and the nuclear surfaces overlap, the expressions for the Coulomb interaction become more complex. However, it is not important for the low-energy processes of heavy-ion collisions considered here, as with $R < R_1 + R_2$ the nuclear system is governed by the adiabatic potential energy (see Chap. 5). For deformed nuclei, their quadrupole and/or hexadecupole deformations are usually taken into account. Since, as a rule, $\beta_4 \ll 1$, in the third part only the terms with $\lambda' = \lambda'' = 2$ remain unchanged and λ takes the values 2 and 4.

The short-range nuclear interaction essentially depends on the distance between the surfaces of the colliding nuclei. In this case the distance along the nuclear separation axis $\xi = R - R_1(\vec{\beta}_1, \theta_1) - R_2(\vec{\beta}_2, \theta_2)$ is generally used (see Fig. 2.4). The interaction is often approximated by the Woods–Saxon potential

$$V_{WS}(\xi) = \frac{V_0}{1 + \exp\left(\frac{\xi}{a_V}\right)}, \tag{2.12}$$

where $\xi = R - R_V - \Delta R_1 - \Delta R_2$, and $\Delta R_1 = R_1(\vec{\beta}_1, \theta_1) - R_1, \Delta R_2 = R_2(\vec{\beta}_2, \theta_2) - R_2$. Note that for the Woods–Saxon potential the interaction radius $R_V = r_0^V(A_1^{1/3} + A_2^{1/3})$ usually does not coincide with the sum of the radii of the nuclei themselves and r_0^V is an independent parameter (in addition to V_0 and a_V). For light nuclei ($A_1 \leq 4$), the same parameterization of the interaction potential radius as for nucleons, namely $R_V = r_0^V A_2^{1/3}$ is frequently used.

2.2.3 Proximity Potential

Another more suitable way to describe the nucleus–nucleus interaction is to use the *proximity potential* [13]:

$$V_{prox}(\xi) = 4\pi \gamma b P_{sph}^{-1} \cdot \Phi(\xi/b). \tag{2.13}$$

Here $\Phi(\xi/b)$ is the generalized dimensionless form factor:

$$\Phi(x) = \begin{cases} -1.7817 + 0.9270x + 0.14300x^2 - 0.09000x^3, x < 0, \\ -1.7817 + 0.9270x + 0.01696x^2 - 0.05148x^3, \ 0 < x < 1.9475, \\ -4.41 \exp\left(-\frac{x}{0.7176}\right), x > 1.9475; \end{cases}$$

$$\tag{2.14}$$

where b is the parameter of the surface layer thickness (≈ 1 fm); $\gamma = \gamma_0(1 - 1.7826 \cdot I^2)$ with $\gamma_0 \approx 0.95$ MeV $\cdot$ fm^{-2} the surface tension factor, $I = (N - Z)/A$, $P_{sph} = 1/\bar{R}_1 + 1/\bar{R}_2$ is the resultant curvature of the surfaces of spherical nuclei and $\bar{R}_i = R_i[1 - (b/R_i)^2]$. This interaction is most sensitive to the choice of nuclear matter radii. Meaningful results are obtained, when we choose $r_0 \approx 1.16$ fm for heavy nuclei radii ($A > 50$) and $r_0 \approx 1.22$ fm for nuclei with $A \sim 16$. For intermediate values of A, we can use the expression $r_0 \approx 1.16 + 16/A^2$, and for $A < 16$, the value of r_0 should be chosen individually (see Fig. 1.2). The main advantage of the proximity potential lies in its universality, i. e., in the absence of parameters to be adjusted, such as V_0, r_0^V, a_V (as for the Woods–Saxon potential).

The magnitude of the attraction between two nuclear surfaces also depends on their curvature, that is, on the area of the contacting surfaces. Usually this is taken into account by replacing P_{sph} in Eq. (2.13) by the expression

$$P(\vec{\beta}_1, \theta_1, \vec{\beta}_2, \theta_2) = \left[(k_1^\parallel + k_2^\parallel)(k_1^\perp + k_2^\perp) \right]^{1/2}, \qquad (2.15)$$

where $k_i^{\parallel,\perp}$ are the main parameters of the local curvature of the projectile and target surfaces. For spherical nuclei $k_i^{\parallel,\perp} = R_i^{-1}$ and $P = P_{sph}$. In the case of interaxial dynamic deformations (face-to-face orientation), $\theta_1 = \theta_2 = 0$, that occur in the slow collisions of dynamically deformed nuclei, the local curvature of the surfaces can be cast in the explicit form:

$$P(\vec{\beta}_1, \theta_1 = 0, \vec{\beta}_2, \theta_2 = 0) = \sum_{i=1,2} \frac{1}{\tilde{R}_i} \left(1 + \sum_{\lambda \geqslant 2} \sqrt{\frac{2\lambda+1}{4\pi}} \beta_{i\lambda} \right)^{-2} \times$$

$$\times \left(1 + \sum_{\lambda \geqslant 2} (1 + \eta(\lambda)) \sqrt{\frac{2\lambda+1}{4\pi}} \beta_{i\lambda} \right), \qquad (2.16)$$

where $\eta(\lambda) = \frac{1}{2}\lambda(\lambda + 1)$.

2.2.4 Bass Potential

To estimate the Coulomb barrier between two sufficiently heavy nuclei, we often use another phenomenological potential known as the Bass potential [7]:

$$V_{Bass}(R) = \frac{Z_1 Z_2 e^2}{R} - \frac{R_1 R_2}{R_1 + R_2} g(\xi), \qquad (2.17)$$

where $\xi = R - (R_1 + R_2)$ is the distance between the nuclear surfaces, nuclear radii are determined by the expression $R_i = 1.16 A_i^{1/3} - 1.39 A_i^{-1/3}$, and the function $g(\xi) = \left[A \exp\left(\frac{\xi}{d_1}\right) + B \exp\left(\frac{\xi}{d_2}\right) \right]^{-1}$ with the parameters $A = 0.03\,\mathrm{MeV}^{-1} \cdot \mathrm{fm}$, $B = 0.0061\,\mathrm{MeV}^{-1} \cdot \mathrm{fm}$, $d_1 = 3.3\,\mathrm{fm}$, and $d_2 = 0.65\,\mathrm{fm}$.

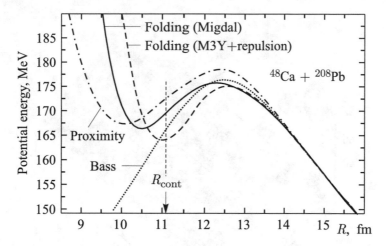

Fig. 2.5 Potential interaction energy of the spherical nuclei ^{48}Ca $+^{208}$Pb. The solid curve is the folding potential with zero-range forces, the dashed curve is folding with the M3Y-interaction plus repulsion (see the parameters given above), and the dotted and dash-dotted curves are the phenomenological Bass [7] and proximity potentials [13], respectively. R_{cont} indicates the distance between the centers of the touching nuclei

2.2.5 Comparison of Diabatic Potentials for the Nucleus–Nucleus Interaction

A comparison of the interaction potentials obtained in the above models is shown in Fig. 2.5 for the collision of the spherical nuclei ^{48}Ca and ^{208}Pb. With properly selected parameters, different models give similar values of Coulomb barriers (within 2 or 3 MeV). However, at small distances (in the region of overlapping nuclear surfaces) these potentials differ greatly. The available experimental data cannot provide a more accurate determination of the diabatic nucleus–nucleus potential at small distances. The main reason for this is a strong connection of the relative motion of the nuclei (evolution of the coordinate R) and many other internal degrees of freedom of the overlapping nuclei. Using the one-dimensional potential $V(R)$ in this region becomes too simplistic, and the evolution of the system should be determined by a multidimensional driving potential (see Chap. 5).

2.2.6 Dependence of Potential Energy on Nuclear Orientation

As noted above, some nuclei are deformed in their ground state. Naturally, their interactions then depend on their spatial orientation (see Fig. 2.4). Short-range attractive nuclear forces depend on the distance between the surfaces of the nuclei $\xi = R - R_1(\vec{\beta}_1, \theta_1) - R_2(\vec{\beta}_2, \theta_2)$, whereas long-range Coulomb repulsive forces

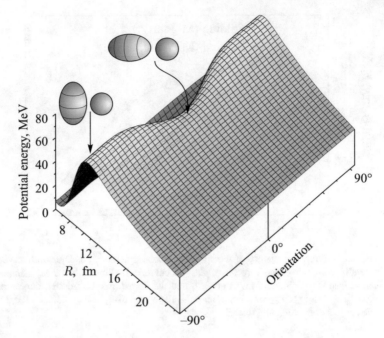

Fig. 2.6 Potential energy of the nuclear interaction of ^{16}O + ^{154}Sm ($\beta_2 = 0.3$, $\beta_4 = 0.1$) as a function of distance and orientation

depend on the distance between the nuclear centers R. Therefore, if, for example, we consider the configuration of two touching nuclei ($\xi = 0$), the nuclear forces remain constant and the Coulomb energy decreases sharply for *face-to-face* orientation, as in this case the distance between the centers of the nuclei grows. As a result, for this configuration, the total potential energy decreases as well. The interaction of the spherical nucleus ^{16}O with the deformed nucleus ^{154}Sm ($\beta_2 = 0.3$, $\beta_4 = 0.1$) is shown in Fig. 2.6 as a function of the ^{154}Sm orientation. It can be seen that the height of the Coulomb barrier strongly depends on the orientation of the deformed nucleus—the difference for limit orientations (*side-to-side* and *face-to-face*) can exceed 20 MeV for heavy statically deformed nuclei.

The azimuthal orientation of two deformed nuclei affects the potential energy of their interaction much less. Figure 2.7 shows the potential energy of two statically deformed nuclei, ^{64}Zn ($\beta_2 = 0.22$) and ^{150}Nd ($\beta_2 = 0.24$), depending on their polar and azimuthal orientation (rotation in a plane perpendicular to the internuclear axis, see Fig. 2.4. As can be seen from the figure, the Coulomb barrier changes only by 1 or 2 MeV with a change in the azimuthal orientation of the nuclei (irrespective of their polar orientation), and thus such rotation can be neglected. A more detailed description of the potential energy for heavy nuclei interactions, as well as algorithms for its calculation can be found in [78] and in the nuclear knowledge base on the website http://nrv.jinr.ru/nrv.

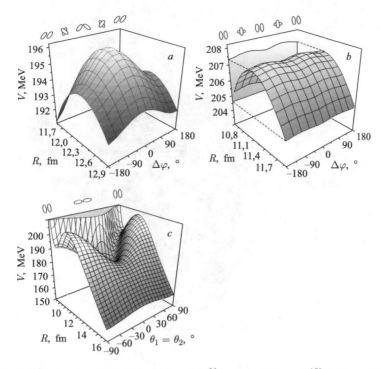

Fig. 2.7 Potential energy of two deformed nuclei, ^{64}Zn($\beta_2 = 0.22$) and ^{150}Nd($\beta_2 = 0.24$) as a function of their azimuthal (**a** and **b**) and polar (**c**) orientation. (**a**) $\theta_1 = \theta_2 = \pi/4$; (**b**) $\theta_1 = \theta_2 = \pi/2$; (**c**) $\varphi_1 = \varphi_2$, i.e., $\Delta\varphi = 0$. A schematic view of nuclear orientation is shown at the top of the figures

2.2.7 Dependence of Potential Energy on Dynamical Deformations

In collisions, heavy nuclei that are spherical in their ground states can experience dynamic deformations. Figure 2.8 shows the potential interaction energy for the nuclei of ^{40}Ca and ^{90}Zr as a function of their dynamic quadrupole deformations, calculated using the parameters of the liquid-drop model. For simplicity, it was assumed that the nuclear deformation energy is proportional to the mass, and instead of the two dynamical deformation parameters β_1 and β_2 only one was used: $\beta = \beta_1 + \beta_2$ (these indices refer to the first and second nucleus; in both cases the deformation is quadrupole). It is seen from the figure that the height of the Coulomb barrier strongly depends on the dynamic deformation of the nuclei. In the region of the Coulomb barrier, i.e., at distances $R > R_1 + R_2$, an increase in the deformation leads to a decrease in the distance between the surfaces of the nuclei and an increase in the force of attraction. With a further increase in dynamic deformation, the potential energy begins to rise due to elasticity of the nuclear surface, since the energy of deformation should be added to the value of the internuclear interaction

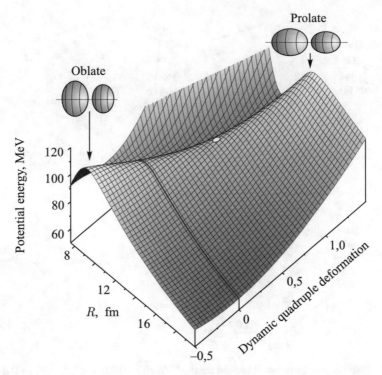

Fig. 2.8 Potential interaction energy for the ^{40}Ca + ^{90}Zr system (both nuclei spherical in the ground state) as a function of the distance between their centers of mass and dynamic deformation. The minimum value of the Coulomb barrier height (marked by a circle) is achieved at a nonzero dynamic deformation. The red line on the surface indicates the corresponding potential for spherical nuclei

$V_{12}(R; \beta_1, \beta_2)$, which at small values of β_1 and β_2 has the form $\frac{1}{2}C_1\beta_1^2 + \frac{1}{2}C_2\beta_2^2$, where $C_{1,2}$ are the rigidities of the nuclei surfaces. As a result, the Coulomb barrier takes the form of a ridge with a saddle (minimum) point at a certain nonzero value of dynamic deformation. As we shall see below, it is this behavior of the Coulomb barrier that leads to an increase in the nuclear fusion cross section at sub-barrier energies.

2.3 Classification of Nuclear Reactions, Experimental Procedures, Cross Sections, and Kinematics

Nuclear reactions with heavy ions are used for a wide variety of purposes. With their help we can study the properties of nuclei, obtain and investigate exotic nuclear states (highly excited, rapidly rotating, etc.) and nuclei at the drip lines (proton- and neutron-rich), synthesize superheavy nuclei (including new chemical elements), explore nuclear reactions mechanisms, and so on. Depending on the desired goal, a suitable collision energy, the required combination of target and projectile nuclei and the corresponding reaction mechanism are chosen. Collision energies can be roughly divided into *low* (from zero to 150 MeV/nucleon, that is, up to the threshold for the production of the lightest mesons), *intermediate* (from 150 to 1000 MeV/nucleon, that is, up to the baryon production threshold), *high* (from 1 to 100 GeV/nucleon), and *superhigh* (several TeV/nucleon). Needless to say, this classification is very rough. But in this course we are interested in low collision energies, at which no new particles are created (or their production is extremely unlikely) and the number of nucleons remains unchanged.

However, in low-energy collisions of heavy ions, one can still observe a large number of different processes: elastic scattering of the colliding particles, quasi-elastic scattering and few-nucleon transfer, deep-inelastic scattering, fusion reactions, and fragmentation and fission processes. In the following chapters, these different nuclear reaction mechanisms are considered.

The simplest experimental procedure is shown in Fig. 2.9a. Here a beam of monochromatic ions from an accelerator bombards a target. At a distance d from the target there is a detector with a window area ΔS, covering a solid angle $\Delta \Omega = \Delta S / d^2$. Let us denote the cross section of the nucleus by σ (see Fig. 2.9b). If the beam of accelerated particles strikes a target of area S and the thickness of the target is ℓ, then the visible area of the target nuclei is $\Sigma = \sigma \cdot \rho \cdot S \cdot \ell$, where ρ is the concentration of atoms in the target material. The probability that one of the bombarding particles hits a target nucleus is $P = \Sigma / S = \sigma \cdot \rho \cdot \ell$, and the number of events (reactions) per second $N = \nu_0 \cdot \sigma \cdot \rho \cdot \ell$ also depends on the intensity of the beam ν_0 [particles per second]. Let us denote that part of the total cross section that leads to scattering at an angle θ into the solid angle $\Delta \Omega$ by $\Delta \sigma (\theta)$. Then the count rate of the detector is $\Delta N = \nu_0 \cdot \Delta \sigma \cdot \rho \cdot \ell$ [events per second]. Obviously, the count rate is directly proportional to the detector area ΔS (for sufficiently large distances). Therefore, it is convenient to determine the so-called differential cross section:

$$\frac{d\sigma}{d\Omega}(\theta) = \frac{\Delta N(\theta)}{\nu_0 \rho \ell \Delta \Omega}. \tag{2.18}$$

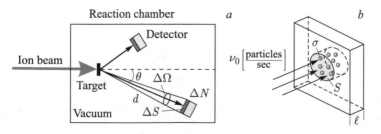

Fig. 2.9 Experimental procedure (**a**) and determination of the nuclear reaction cross section (**b**)

This value does not depend on the intensity of the beam, the density and thickness of the target, or the properties of the detector. It reflects only the nature of the interaction of the projectile particles with the atomic nuclei of the target material. In more complex experiments, detectors not only register the fact that a particle passes through, but also measures its energy or registers the fact that it passes through in coincidence with another particle passing through another detector (for example, in the process of fragmentation). In this case multi-dimensional cross sections are measured: $d^2\sigma/d\Omega dE$, $d^3\sigma/d\Omega_1 d\Omega_2 dE_2$, etc. (see below).

Calculations and the analyses of collision processes should be carried out in the center-of-mass system, while measurements of course are made in the laboratory system. It becomes necessary, therefore, to transform data from one system to the other. This is done using kinematic ratios based on the laws of conservation of energy and momentum. Let us suppose, for example, that a nucleus with mass number A_1 and velocity v_{lab} hits a stationary target nucleus of mass A_2 and, as a result of the collision, nuclei with masses A_3 and A_4 are formed (Fig. 2.10a). It is obvious that $E_{\text{lab}} = \frac{m_N A_1}{2} v_{\text{lab}}^2$, $E_{\text{c.m}} = \frac{A_2}{A_1+A_2} E_{\text{lab}} = \frac{\mu}{2} v_{\text{lab}}^2$, where $\mu = \frac{A_1 A_2}{A_1+A_2} m_N$ is the reduced mass of the system and m_N is the nucleon mass. In the center-of-mass system, the colliding particles move towards one other with velocities $v_1 = \frac{A_2}{A_1+A_2} v_{\text{lab}}$ and $v_2 = \frac{A_1}{A_1+A_2} v_{\text{lab}}$, and their momenta are equal: $A_1 v_1 = A_2 v_2$. The momenta of the resulting particles (at the angles θ and $\pi - \theta$) are also equal: $A_3 v_3 = A_4 v_4$. The entire collision process in the center-of-mass system is illustrated in Fig. 2.10b. The center-of-mass of the system moves in the laboratory frame with velocity $U_{c.m.} = \frac{A_1 v_{lab} + A_2 v_{lab}^{(2)}}{A_1+A_2} = \frac{A_1}{A_1+A_2} v_{lab}$, so when we change from the laboratory system we should add $\mathbf{U}_{c.m.}$ to the velocity of each particle $\mathbf{v}_i$ (Fig. 2.10c). As a result, the scattering angles in the laboratory system are smaller than those in the center-of-mass frame. For fragment 3 shown in Fig. 2.10 these angles are related as follows

$$\tan\theta_3^{lab} = \frac{v_3 \sin\theta}{v_3 \cos\theta + U_{c.m.}} = \frac{\sin\theta}{\cos\theta + \frac{U_{c.m.}}{v_3}}. \qquad (2.19)$$

In the case of elastic scattering $v_3 = v_1 = \frac{A_2}{A_1+A_2} v_{lab}$, $\frac{U_{c.m.}}{v_3} = \frac{A_1}{A_2}$, and $\tan\theta_3^{lab} = \frac{\sin\theta}{\cos\theta + \frac{A_1}{A_2}}$. For identical-particle collisions $(A_1 = A_2)$ $\theta^{lab} = \theta/2$.

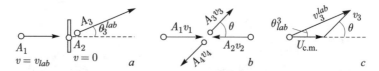

Fig. 2.10 Kinematics of the simplest nuclear reaction. (**a**) Experimental setup. (**b**) Center-of-mass system. (**c**) Laboratory system

Conversion of the differential cross section from the center-of-mass system to the laboratory frame (or vice versa) is based on the invariance of the number of particles ΔN hitting the detector in both coordinate systems, that is, $\frac{d\sigma^{lab}}{d\Omega}(\theta^{lab}) = \frac{d\sigma^{c.m.}}{d\Omega}(\theta)\frac{\sin\theta}{\sin\theta^{lab}}\frac{d\theta}{d\theta^{lab}}$. In particular, for elastic scattering we obtain the following formula to convert the differential cross sections:

$$\frac{d\sigma^{lab}}{d\Omega}(\theta^{lab}) = \frac{x\sqrt{x}}{|1+\alpha\cos\theta|}\frac{d\sigma^{c.m.}}{d\Omega}(\theta), \qquad (2.20)$$

where $\alpha = A_1/A_2$, $x = 1 + \alpha^2 + 2\alpha\cos\theta$, and the angles θ and θ^{lab} are related by the ratio (2.19).

The laws of conservation of total energy and momentum give four relations altogether. In the case of a two-body reaction $A_1 + A_2 \rightarrow A_3 + A_4$ (at low energies, it is mainly these processes that occur) out of six unknowns (v_3^x, v_3^y, v_3^z, v_4^x, v_4^y, v_4^z or $E_3, \theta_3, \varphi_3, E_4, \theta_4, \varphi_4$) two variables remain arbitrary. One of them is fixed by choosing the reaction plane passing through the beam axis and the line connecting the target and the detector (see Fig. 2.9). Clearly, due to azimuthal symmetry the choice of this plane is arbitrary and based on the convenience of detector location. If we denote this plane by (y, z), then $v_3^x = 0$ or $\varphi_3 = 0$ and there are five remaining variables with four equations defining the laws of conservation. This means that in two-body reaction channels there is only one value left to be measured, and all the remaining ones can be easily determined from the corresponding conservation laws. It is usually the emission angle of one of the fragments that we choose for this purpose. Hence in the experiment one simply measures the number of particles registered by the detector, depending on its rotation angle. Afterwards, this number is converted into a *one-dimensional* differential cross section by the formula (2.18). In this case, it is not necessary to measure the energy of this fragment as well as the emission angle and the energy of the second fragment. They can be easily and unambiguously determined (in the center-of-mass system) from the conservation laws. As an example, Fig. 2.11 shows possible energies and emission angles in the laboratory system of ^{17}F and ^{12}B nuclei produced in the reaction ^{20}Ne $+$ ^{9}Be at an energy of 50 MeV/nucleon. It is clear that in order to study this reaction, one detector (but not two, as shown for clarity in Fig. 2.11) is sufficient. The situation does not change, if the emitted nucleus is formed in one of its excited states, for example, $A_1 + A_2 \rightarrow A_3'(\varepsilon_n) + A_4$. In this case, the excitation energy ε_n should simply be taken into account in the law of total energy conservation, whereas the detector should be able to distinguish between the excited state of the nucleus A_3 and its ground state. In this case we measure the differential cross section $d\sigma^{(n)}(\theta)/d\Omega$ in the channel n.

However, in the collision of sufficiently heavy nuclei, fragments with high excitation energies can be produced in the emission channel (see deep-inelastic scattering reactions below). In this case, the excitation energy of both fragments (that is, the loss of kinetic energy) is another variable, and it is of great interest to measure the double differential section $d^2\sigma/d\Omega dE$ even for a two-body reaction.

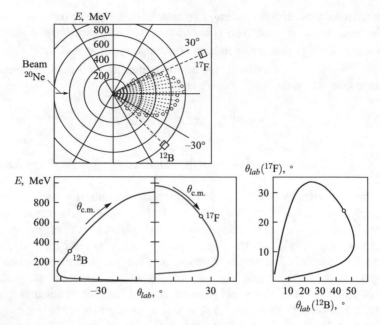

Fig. 2.11 Kinematics of a two-body nuclear reaction in the laboratory coordinate system. The figure shows ^{20}Ne colliding with a beryllium target at an energy of 50 MeV/nucleon. As a result, nuclei of ^{17}F and ^{12}B, in their ground state, are produced in the emission channel. Measuring, for example, the energy of one of the emitted nuclei uniquely determines the angle and energy of the second nucleus

As the collision energy increases, it is highly probable that more than two fragments will be produced in the emission channel and the situation will change radically. In reactions involving weakly bound nuclei (for example, ^{2}H, ^{6}He, etc.) such processes also occur at low collision energies. Since there are still only four conservation laws, the angles of emission and the energies of the particles formed can take arbitrary values in certain predetermined intervals. As an example, Fig. 2.12 shows the kinematics of the three-body reaction ^{20}Ne $+ {}^{9}$Be $\rightarrow {}^{17}$F $+ {}^{9}$Be $+ {}^{3}$H at a beam energy of 50 MeV/nucleon. Any point inside the shaded regions determines the emission energy and angle of the corresponding fragment, which can be obtained in one of the recorded events. As can be seen from the figure, these values are not entirely arbitrary. The conservation laws impose certain restrictions on the emission energy and angles of all particles. These laws, however, do not allow us to conclude what energies and at what angles there will be more or less particles emitted (for this purpose one must solve the dynamical equations of motion, taking into account the particle interactions). To study the mechanism of such a reaction in full, one must measure the emission angle of two particles and the energy of one of them (or two energies and one angle), that is, using two independent detectors.

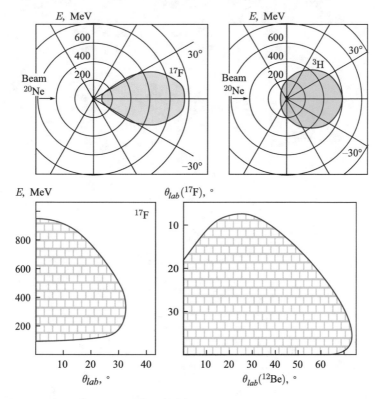

Fig. 2.12 Kinematics of a three-body nuclear reaction in the laboratory coordinate system. The figure shows ^{20}Ne colliding with a beryllium target at an energy of 50 MeV/nucleon. As a result, nuclei of ^{17}F, ^{9}B, and ^{3}H are produced in the emission channel. The emitted nuclei may have any angle and energy values in the corresponding shaded regions

Chapter 3
Elastic Scattering of Nucleons and Heavy Ions

3.1 Scattering in a Coulomb Field

At low collision energies, the Coulomb potential between two nuclei plays an important role. The attractive nuclear forces (which begin to act when the distance between the two surfaces is around 1–2 fm) and the repulsive Coulomb field provide the total interaction and give rise to a Coulomb barrier of height V_B at a distance $R_B > R_{cont} = R_1 + R_2$ (see Fig. 3.1a). At energies below that height (exemplified by E_1 in Fig. 3.1), the colliding nuclei do not make close contact. The nuclear forces then play little role and the only reaction channels observed are elastic scattering in the Coulomb field $V(r) = Z_1 Z_2 e^2 / r$ along with a small probability of inelastic excitations of the target and projectile nuclei (see below).

The field of trajectories for sub-barrier collisions of ^{48}Ca nuclei with ^{208}Pb nuclei is shown in Fig. 3.1b. It can be seen that even in head-on collisions ($b = 0$), the nuclei cannot overcome the Coulomb barrier—the shaded region is "classically forbidden"; the wavefunction describing the relative motion of the nuclei decreases exponentially in this region (see below). A three-dimensional surface that separates the accessible and inaccessible areas of motion, $\theta_C(r)$, is called the caustic surface. It is represented by the dashed curve in Fig. 3.1b. In the accessible region of motion, two Coulomb trajectories pass through each point of the (r, θ) space (bold curves in Fig. 3.1b); they have impact parameters b_2 (that has not yet reached the caustic surface) and b_1 (that has already been reflected). These satisfy

$$b_{1,2}(r, \theta) = \frac{r \sin \theta}{2} \left\{ 1 \pm \sqrt{1 - \frac{4\eta}{kr} \frac{1}{(1 - \cos \theta)}} \right\}. \tag{3.1}$$

At a distance r from the scattering center, the deflection angle of the Coulomb trajectory with impact parameter b (the distance from the beam axis at large r) is

© Springer Nature Switzerland AG 2019
V. Zagrebaev, *Heavy Ion Reactions at Low Energies*, Lecture Notes in Physics 963,
https://doi.org/10.1007/978-3-030-27217-3_3

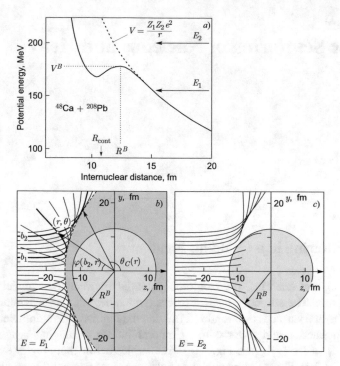

Fig. 3.1 (**a**) Potential energy of interaction for scattering of ^{48}Ca nuclei by ^{208}Pb. (**b**) The field of trajectories at the sub-barrier energy $E_1 = 160$ MeV and (**c**) above the barrier at an energy $E_2 = 200$ MeV in the center-of-mass system

given by the formula

$$\varphi_C(b,r) = \arctan \frac{kb/r + \eta/b}{k(b,r)} - \arctan \frac{\eta}{kb}, \qquad (3.2)$$

where $k = \sqrt{2mE/\hbar^2}$ is the wave number, $k(b,r) = k\sqrt{1 - \frac{V(r)}{E} - \frac{b^2}{r^2}}$, and $\eta = kZ_1Z_2e^2/2E$ is known as the Sommerfeld parameter. The asymptotic deflection angle (for $r \to \infty$) is given by the expression

$$\theta(b) = \pi - 2\varphi_C(b, r_0) = 2\arctan(\eta/kb), \qquad (3.3)$$

where $r_0(b) = \eta/k + \sqrt{(\eta/k)^2 + b^2}$ is the distance of closest approach for particles with impact parameter b. The caustic surface is defined by the relation $\theta_C(r) = \arccos(1 - 4\eta/kr)$ or $r_C(\theta) = 4\eta/[k(1 - \cos\theta)]$.

In classical mechanics, the differential cross section for elastic scattering is relatively simple. A detector fixed at an angle θ, and having a window that covers the solid angle $\Delta\Omega = \sin\theta \Delta\theta \Delta\phi$, will detect all the particles that move along

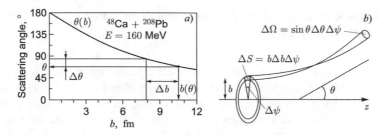

Fig. 3.2 (**a**) The deflection function for sub-barrier scattering of ^{48}Ca nuclei with energy 160 MeV by ^{208}Pb nuclei (Eq. (3.3) with $k = 17.3$ fm^{-1} and $\eta = 127.5$). (**b**) Schematic representation of scattering of particles in classical mechanics

trajectories with impact parameters close to $b(\theta)$. The *tube* of such trajectories is shown in Fig. 3.2b. The number of particles impinging on the detector every second is equal to $\Delta N = n_0 \cdot \Delta S$, where n_0 is the beam flux [particles/cm^2/sec], and $\Delta S = b\Delta b \cdot \Delta \phi$ is the cross sectional area of the trajectory tube (ϕ is the azimuthal angle, see Fig. 3.2). The differential cross section in this case is defined as $\Delta \sigma / \Delta \Omega = \Delta N / n_0 \Delta \Omega$, that is,

$$\frac{d\sigma^{cl}}{d\Omega}(\theta) = \frac{b(\theta)}{\sin \theta} \frac{1}{|d\theta(b)/db|}, \tag{3.4}$$

where $\theta(b)$ is the deflection function and $b(\theta)$ is the corresponding inverse function.

For the case of scattering by a Coulomb field, the deflection angle has the simple form of Eq. (3.3), that is, $b(\theta) = \frac{\eta}{k} \cot(\theta/2)$, and the differential cross section (3.4) reduces to the expression

$$\frac{d\sigma_R}{d\Omega}(\theta) = \left(\frac{Z_1 Z_2 e^2}{4E}\right)^2 \frac{1}{\sin^4\left(\frac{\theta}{2}\right)}. \tag{3.5}$$

This is referred to as the Rutherford formula, and we note that the quantum description of this scattering process (see below) leads to exactly the same expression.

If the collision energy exceeds the height of the Coulomb barrier (for example, the case $E = E_2$ in Fig. 3.1), the incident particles with small impact parameters reach the surface of the target nucleus and their further evolution is influenced by the nuclear attraction. However, for large impact parameters that lead to scattering at small angles, the particles will still move along Coulomb trajectories (see Fig. 3.1c) and consequently the cross section for elastic scattering at small angles is, as before, determined by Rutherford's formula (3.5). This is why, when analyzing experimental data on elastic scattering, the ratio of the experimental differential cross section to the Rutherford cross section $d\sigma/d\sigma_R$ is often considered, since it gives a more vivid graphic representation of the role of nuclear forces.

3.2   Elastic Scattering of Protons and Neutrons: Optical Model

A typical differential cross section for the elastic scattering of protons on atomic nuclei is shown in Fig. 3.3. A characteristic feature of this cross section is oscillations in the angular distribution (interference pattern) reflecting the wave character of proton scattering. An analogous picture is also observed in the case of elastic neutron scattering, see above Fig. 1.3a. This means that in describing the elastic scattering of nucleons by atomic nuclei, we need to use the quantum approach. Indeed, the de Broglie wavelength of the nucleon $\lambda = 2\pi/k = 2\pi/\sqrt{2mE/\hbar^2} \approx 28.6/\sqrt{E}$ (here the energy is measured in MeV) at the low collision energies of interest to us is comparable to or larger than the nucleus size. An exact solution of the many-body quantum mechanical problem of nucleon–nucleus scattering is totally impracticable. Therefore, when describing the elastic scattering of nucleons (and more heavy nuclei), the so-called optical model (OM) is employed.

The optical model of elastic scattering of nuclei was first proposed by Feshbach, Porter, and Weisskopf [26], and its theoretical justification was further elucidated, with the help of so-called projection operators, by Feshbach [24, 25]. The essence of this model is the introduction of a complex potential $V(r)+iW(r)$ (called the optical potential). This describes the relative motion of particles in the elastic channel and also the decrease of their flux due to absorption into inelastic channels (provided by a negative imaginary part W to the potential).

In the collision of particles $a + A$, in addition to their elastic scattering, other processes (referred to as reaction channels) can occur. Examples are the inelastic excitation of the target nucleus $a' + A^*$, or the breakup of the incident particle $b + c + A$, the fusion of the two nuclei $a + A \rightarrow B$, etc. The total Hamiltonian of

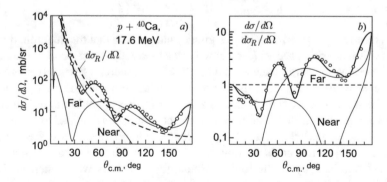

Fig. 3.3 (a) Differential cross section for proton elastic scattering by ^{40}Ca nuclei at an energy of 17.6 MeV [21]. (b) The ratio of the differential cross section to the Rutherford cross section (3.5). Thin curves show the contributions of the so-called *near-side* and *far-side* components to the scattering (see text)

such a system has the form

$$\hat{H}(r, \xi_a, \xi_A) = \hat{t}(r) + \hat{H}_a(\xi_a) + \hat{H}_A(\xi_A) + V_{aA}(r, \xi_a, \xi_A),$$

where $\hat{t} = -\hbar^2 \nabla^2 / 2\mu$ is the kinetic energy operator of the relative motion, and $\hat{H}_a$ and $\hat{H}_A$ are the internal Hamiltonians for the motions within a and A. These are described by the wavefunctions Φ_m^a and Φ_n^A, respectively:

$$\hat{H}_a \left| \Phi_m^a \right\rangle = \varepsilon_m^a \left| \Phi_m^a \right\rangle, \, \hat{H}_A \left| \Phi_n^A \right\rangle = \varepsilon_n^A \left| \Phi_n^A \right\rangle.$$

If a is an unstructured, inert particle, then $\hat{H}_a$ and Φ_m^a can of course be ignored. The term $V_{aA}(r, \xi_a, \xi_A)$ is the potential energy of the interaction between a and A. Using the completeness of the states Φ_m^a and Φ_n^A, one can introduce the projection operators

$$\hat{P} = \left| \Phi_0^a, \Phi_0^A \right\rangle \left\langle \Phi_0^A, \Phi_0^a \right|$$

and

$$\hat{Q} = \sum_{m \neq 0} \sum_{n \neq 0} \left| \Phi_m^a \Phi_n^A \right\rangle \left\langle \Phi_n^A \Phi_m^a \right|.$$

The operators $\hat{P}$ and $\hat{Q}$ satisfy the following conditions

$$\hat{P} + \hat{Q} = \hat{1}, \quad \hat{P} \cdot \hat{Q} = 0$$

because of the orthogonality of these states: $\left\langle \Phi_m^a \mid \Phi_{m'}^a \right\rangle = \delta_{mm'}$ and $\left\langle \Phi_n^A \mid \Phi_{n'}^A \right\rangle = \delta_{nn'}$. The projection of the total wavefunction onto the ground state of the projectile and target

$$\left\langle \Phi_0^A \Phi_0^a \mid \Psi_{\vec{k}}^{(+)}(a, A) \right\rangle \equiv \left| \psi_{\vec{k}}^{(+)} \right\rangle$$

clearly describes the relative motion of particles with momentum $\vec{k}$ exactly in the elastic channel (the sign $(+)$ signifies the choice of boundary conditions with a divergent scattered wave at large distances).

The Schrödinger equation for the total wavefunction

$$\hat{H} \left| \Psi_{\vec{k}}^{(+)} \right\rangle = E \left| \Psi_{\vec{k}}^{(+)} \right\rangle$$

can easily be divided into a system of two equations

$$\hat{P} \hat{H} \hat{P} + \hat{P} \hat{H} \hat{Q} \left| \Psi \right\rangle = E \hat{P} \left| \Psi \right\rangle$$

and

$$\hat{Q}\hat{H}\hat{Q} + \hat{Q}\hat{H}\hat{P} |\Psi\rangle = E\hat{Q} |\Psi\rangle .$$

One can thus obtain an effective equation for the function $\left| \psi_{\vec{k}}^{(+)} \right\rangle$ that describes the process of elastic scattering

$$\left[\hat{t}_r + \hat{P} V_{aA} \hat{P} + \hat{P} V_{aA} \hat{Q} \frac{1}{E + i0 + \hat{Q}\hat{H}\hat{Q}} \hat{Q} V_{aA} \hat{P} \right] \left| \psi_{\vec{k}}^{(+)} \right\rangle$$

$$= (E - \varepsilon_0^a - \varepsilon_0^A) \left| \psi_{\vec{k}}^{(+)} \right\rangle . \qquad (3.6)$$

The generalized optical potential obtained in this way (the sum of the second and third terms in square brackets)

$$V_{aA}^{OM} = \left\langle \Phi_0^A \Phi_0^a \left| V_{aA} \right| \Phi_0^a \Phi_0^A \right\rangle$$

$$+ \sum_{m\neq 0, n\neq 0} \left\langle \Phi_0^A \Phi_0^a \left| V_{aA} \right| \Phi_m^a \Phi_m^A \right\rangle \frac{1}{E + i0 + \hat{Q}\hat{H}\hat{Q}} \left\langle \Phi_m^A \Phi_n^a \left| V_{aA} \right| \Phi_0^a \Phi_0^A \right\rangle ,$$

$$\qquad (3.7)$$

has a fairly transparent structure. The first term is an ordinary folding potential (see the previous chapter), that is, the interaction of the nucleons comprising a and A, averaged over the internal motion in their ground states. The second term reflects the coupling of the elastic channel to the reaction channels—due to the interaction V_{aA}, the nuclei can make transitions from their ground states (Φ_0^a, Φ_0^A) to excited ones, $\Phi_{m\neq 0}^a$, $\Phi_{n\neq 0}^A$ and stay there for some time (described by the propagator $[E + i0 + \hat{Q}\hat{H}\hat{Q}]^{-1}$) before returning to the elastic channel. This term contains both real (a polarization term to be added to the folding potential) and imaginary parts. The exact calculation of this term is rather complicated and has only been performed for certain reaction channels (for example, for the breakup channel for the deuteron or for inelastic excitations of vibrational target nuclei).

Generally in the standard optical model a phenomenological potential $U(r; E)$ is used in which the radial dependence and the magnitude of the real and imaginary parts are selected from the analysis of the experimental data on elastic scattering for a given combination of nuclei at a given energy. In the case of nucleon scattering, several global parameterizations have been proposed which satisfactorily describe the experimental data over a fairly wide range of nuclei and energies. Becchetti and Greenlees [9] were the first to propose such parameterizations

$$U(r, E) = V_C(r) - V_V(r, E) - i W_V(r; E) + i W_D(r, E)$$

$$+ (V_{SO}(r, E) + i W_{SO}(r, E)) (\vec{l} \cdot \vec{\sigma}) . \qquad (3.8)$$

Here $V_{V,SO}$ and $W_{V,D,SO}$ are the real and imaginary components of the volume (V), surface (D), and spin-orbital (SO) potentials, and E is the energy of the incident nucleon. All components of the potential are represented in the form of the product of an E-dependent depth, V_V, W_V, W_D, and V_{SO}, and a radial form factor $f(r)$:

$$V_V(r, E) = V_V(E) f(r, R_V, a_V), \tag{3.9}$$

$$W_V(r, E) = W_V(E) f(r, R_W, a_W), \tag{3.10}$$

$$W_D(r, E) = -4a_W W_D(E) \frac{d}{dr} f(r, R_W, a_W), \tag{3.11}$$

$$V_{SO}(r, E) = V_{SO}(E) \left(\frac{\hbar}{m_\pi c}\right)^2 \frac{1}{r} \frac{d}{dr} f(r, R_{SO}, a_{SO}). \tag{3.12}$$

The radial dependence of the form factor $f(r)$ in this case is chosen to have a Woods–Saxon form $f(r) = (1 + \exp[(r - R_i)/a_i])$, where $i = V, W, SO$ and $R_i = r_0^i A^{1/3}$. The Coulomb potential $V_C(r)$ (for protons) is chosen as the interaction of a point charge with a uniformly charged sphere with radius R_C

$$V_C(r) = \begin{cases} \frac{Ze^2}{2R_C}\left(3 - \frac{r^2}{R_C^2}\right), & \text{for } r \leq R_C, \\ \frac{Ze^2}{r}, & \text{for } r > R_C, \end{cases} \tag{3.13}$$

where Z is the charge of the target nucleus. The optical potential parameters (3.9–3.12), acceptable for describing the elastic scattering of nucleons with energy $E \leq 50$ MeV on nuclei with $A > 40$, were selected in [9] and have the values listed in Table 3.1.

Table 3.1 Global optical-model parameters for protons and neutrons induced reactions

Parameters	Protons	Neutrons
V_V (MeV)	$54 + 24\frac{N-Z}{A} - 0.32E + 0.4\frac{Z}{A^{1/3}}$	$56.3 - 24\frac{N-Z}{A} - 0.32E$
R_V (fm)	$1.17A^{1/3}$	$1.17A^{1/3}$
a_V (fm)	0.75	0.75
W_V (MeV)	$0.22E - 2.7$	$0.22E - 1.56$
W_D (MeV)	$11.8 - 0.25E + 12\frac{N-Z}{A}$	$13 - 0.25E - 12\frac{N-Z}{A}$
R_W (fm)	$1.32A^{1/3}$	$1.26A^{1/3}$
a_W (fm)	$0.51 + 0.7(N - Z)/A$	0.58
R_C (fm)	$1.32A^{1/3}$	
V_{SO} (MeV)	6.2	6.2
R_{SO} (fm)	$1.01A^{1/3}$	$1.01A^{1/3}$
a_{SO} (fm)	0.75	0.75

Other variants (including more modern ones) of the global parametrization of nucleon–nucleus optical potentials and light-ion beam OPs can be found in the low-energy nuclear physics knowledge base available on the internet nrv.jinr.ru/nrv/.

The differential cross section for elastic scattering in the framework of the optical model is calculated according to the following scheme. The wavefunction $\psi_{\vec{k}}^{(+)}(\vec{r})$ decomposes into partial waves (for the sake of clarity, we neglect the spin of the particle)

$$\psi_{\vec{k}}^{(+)}(\vec{r}) = 4\pi \sum_{l=0}^{\infty} \sum_{m=-l}^{l} i^l e^{i\sigma_l} \psi_l(k,r) Y_{lm}(\Omega_r) Y_{lm}^*(\Omega_k)$$

$$= \sum_{l=0}^{\infty} (2l+1) i^l e^{i\sigma_l} \psi_l(k,r) P_l(\cos\theta). \tag{3.14}$$

Here $Y_{lm}(\Omega_r)$ are the spherical harmonic functions, the solid angle $\Omega_r \equiv (\theta, \varphi)$ represents the angular variables of the vector $\vec{r}$, where θ is the scattering angle between the vectors $\vec{r}$ and $\vec{k}$, and $\sigma_l = \arg\Gamma(l+1+i\eta)$ are the Coulomb phase shifts (0 for neutrons). Partial wavefunctions are found from the Schrodinger equation with the optical potential $U(r)$

$$\left[\frac{d^2}{dr^2} + \frac{2}{r}\frac{d}{dr} + k^2 - \frac{l(l+1)}{r^2} - \frac{2m}{\hbar^2} U(r) \right] \psi_l(k,r) = 0. \tag{3.15}$$

At large distances (where there are no nuclear forces) these wavefunctions have asymptotic behaviors $\psi_l(r \geq R_M) = \frac{i}{2} e^{i\sigma_l} \left[H_l^{(-)} - S_l H_l^{(+)} \right]$, where $R_M \approx \max(R_V, R_W) + 15 \cdot \max(a_V, a_W)$ is the distance at which the nuclear forces can be neglected, $H_l^{(\pm)}$ are the Coulomb functions, and $S_l = \exp(2i\delta_l)$ are the elements of the so-called scattering S-matrix. They are found from the condition of continuity of the wavefunction and its derivative at the point R_M, that is, from the solution of the algebraic equation

$$\frac{\psi_l'(num)}{\psi_l(num)}\bigg|_{R_M} = \frac{H_l^{'(-)}(kR_M) - S_l H_l^{'(+)}(kR_M)}{H_l^{(-)}(kR_M) - S_l H_l^{(+)}(kR_M)}, \tag{3.16}$$

in which the left-hand side is the result of the numerical integration of equation (3.15) from $r = 0$ to $r = R_M$.

After the partial amplitude $f_l = 2\pi(S_l - 1)/ik$ is found, the differential cross section of elastic scattering is calculated from the formula (in which θ is the center-of-mass scattering angle)

$$\frac{d\sigma}{d\Omega}(\theta) = |f_C(\theta) + f_N(\theta)|^2, \tag{3.17}$$

where the Coulomb and nuclear scattering amplitudes are defined as follows

$$f_C(\theta) = -\frac{\eta}{2k} \frac{1}{\sin^2 \frac{\theta}{2}} \exp\left[2i\left(\sigma_0 - \eta \ln \sin \frac{\theta}{2}\right)\right], \tag{3.18}$$

$$f_N(\theta) = \sum_{l=0}^{\infty} (2l+1)e^{2i\sigma_l} \frac{f_l}{4\pi} P_l(\cos\theta)$$

$$= \frac{1}{2ik} \sum_{l=0}^{\infty} (2l+1)e^{2i\sigma_l}(e^{2i\delta_l} - 1)P_l(\cos\theta). \tag{3.19}$$

Here σ_l are the Coulomb phase shifts and δ_l are the nuclear scattering phases. For a complex optical potential $U(r)$, the phases δ_l are also complex with a positive imaginary part, that is, $|S_l| = |\exp(2i\delta_l)| < 1$ for small l values. For large partial waves, the centrifugal term in the Schrodinger equation (3.15) prevents entry to the small-r region and the partial wavefunction $\psi_l(r \leq R_V) \to 0$. In other words, particles with large impact parameters do not come within range of the nuclear forces (see Fig. 3.1c) and partial amplitudes $f_{l\to\infty} \to 0$ (or $\delta_{l\to\infty} \to 0$, $S_l \to 1$). Thus, the summation in (3.19) needs to be carried out only up to some $l_{max} < kR_M$. For the scattering of nucleons with low energies, there remain just a few low-l terms, while for the scattering of heavy ions, the number of partial waves contributing significantly to the cross section can reach several hundreds.

3.3 Elastic Scattering of Light Ions

The choice of the parameters of the optical potential describing the elastic scattering of light ions is more difficult. In addition to the so-called continuous ambiguity of the parameters (an increase in depth with a compensating decrease in radius, or vice versa), a discrete ambiguity also exists here. If the depth of the nucleon–nucleus potential varies within a few MeV around the value of 50 MeV (the depth of the mean field that reproduces the energies of single-particle bound states), then the depth of the interaction potential, for example for ^{6}Li and ^{12}C, cannot be obtained from the same considerations. This is because we do not know the spectrum of the associated *single-particle* states of these nuclei (there are no such states in pure form because of the multi-nucleon structure of both nuclei).

The correctness of the choice of the OP parameters can be judged by the similarity of the calculated and experimental differential cross sections for elastic scattering. The quantitative characteristic of this fitting is reflected by the quantity

$$\chi^2 = \frac{1}{N} \sum_{i=1}^{N} \frac{\left[\sigma_i^{(th)} - \sigma_i^{(ex)}\right]^2}{\left[\delta\sigma_i^{(ex)}\right]^2}, \tag{3.20}$$

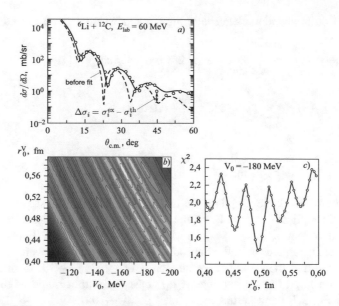

Fig. 3.4 (**a**) Differential cross section for elastic scattering of lithium nuclei on carbon. The experimental data were taken from Ref. [12]. The dotted and solid curves show the result of the calculation with the initial values of the OP parameters and after their automatic adjustment, using standard OM computer codes. (**b**) The topographic landscape of the value of χ^2 in the OP *radius-depth* plane. (**c**) A change in the value χ^2 as a function of the radius of the potential (here $R_V = r_0^V[A_1^{1/3} + A_2^{1/3}]$), when its depth is fixed

that is, the sum over experimental points of the squared deviations of the theoretical and experimental cross sections (see Fig. 3.4), taking into account the errors $\delta\sigma_i^{(ex)}$ in the experimental measurements. This value is a function of the parameters of the OP and its minimization theoretically allows us to find the most appropriate values of these parameters. Repeated calculations of the differential cross section for elastic scattering $\sigma_i^{(th)} \equiv \frac{d\sigma}{d\Omega}(\theta_i)$ is a rather effort-consuming task, and the minimization of χ^2 is usually performed automatically using special computer codes (see, for example, the optical model of elastic scattering in the NRV knowledge base nrv.jinr.ru/nrv). Care is required with this procedure, since the quantity χ^2 may have many local minima (Fig. 3.4).

A distinctive feature of the elastic scattering of light ions of comparable masses is the possibility of few-nucleon transfer to the ground state of the lighter partner. This leads to the incident nucleus taking on the identity of the target, and vice versa. The process is manifested experimentally by an increase in the elastic scattering cross section at backward angles. An example of such an *elastic* transfer is the reaction $^4\text{He} + {}^6\text{Li}(={}^4\text{He} + \text{d}) \rightarrow {}^6\text{Li} + {}^4\text{He}$, in which a deuteron from a target nucleus is transferred to an incident projectile with the formation of a ^6Li nucleus in its ground state. In this case, the newly formed ^4He in the center-of-mass system emerges in the backward direction. As a result, the experimental cross section for

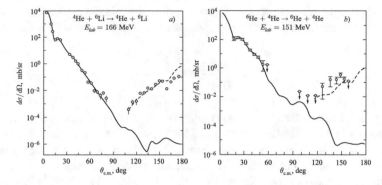

Fig. 3.5 (**a**) Elastic scattering of ^{4}He on ^{6}Li at $E_{lab} = 166$ MeV [5]. (**b**) Elastic scattering of ^{6}He on ^{4}He at energy $E_{lab} = 151$ MeV [66]. The solid curves show the cross sections for elastic scattering calculated within the optical model, and the dashed curves are the contribution of the *elastic transfer* of the deuteron and the di-neutron, respectively

the elastic scattering of ^{4}He on ^{6}Li [5] appears as shown in Fig. 3.5a. Of course the optical model cannot describe the resulting rise in the cross section at backward angles, while the calculation of the corresponding deuteron-transfer cross section, performed in the framework of the distorted-wave method (see below), is capable of explaining this behavior.

In a similar reaction of elastic scattering of the radioactive nucleus ^{6}He on ^{4}He [66], a rise in the cross section at backward angles has also been detected. Again this cannot be explained by any reasonable choice of the OP parameters (see Fig. 3.5b). In this case, this rise is due to the *elastic* transfer of two neutrons: ^{6}He($=^4$He + 2n) + ^{4}He → ^{4}He +^{6}He. A careful analysis of the reaction led to the discovery of a di-neutron state in the nucleus ^{6}He (two spatially correlated neutrons at a considerable distance from the alpha particle). Note, however, that the search for *free* multi-neutron states has not yet been successful.

3.4 Applicability of Classical Mechanics and Trajectory Analyses

In collisions of heavy nuclei (as well as in scattering of nucleons and light ions with sufficiently high energy), the de Broglie wavelength turns out to be sufficiently small to use the regularities, clarity, and language of classical mechanics in the analysis of experimental data. The main advantage of this approach lies not in simpler calculations but in the greater visibility acquired through the use of traditional trajectory analysis. Furthermore, the application of the quasiclassical approximation allows us to take into account quantum mechanical effects such as interference, tunneling, etc.

The de Broglie wavelength $\lambda = \hbar/\mu v = 1/k$ of the relative motion of two nuclei $A_1 + A_2$ with energy E in the center-of-mass system is calculated as follows:

$$\lambda = \sqrt{\frac{\hbar^2}{2\mu E}} \approx \sqrt{\frac{21}{\mu_A E}},$$

where μ is the reduced mass, $\mu_A = A_1 A_2/(A_1 + A_2)$, and the energy $E = \hbar^2 k^2/2\mu$ is measured in MeV ($\hbar^2/2m_N \approx 20.736$ MeV fm^2). Thus, even in the scattering of nucleons with an energy of more than 20 MeV, it is quite possible to use the quasiclassical approximation when calculating wavefunctions and transition amplitudes. For heavier ions, the applicability of a classical analysis is justified for all above-barrier energies. In fact, it is more correct to speak of a local wave number $k(r) = \sqrt{\frac{2\mu}{\hbar^2}[E - V(r)]}$ and a local wavelength $\lambda(r) = 1/k(r)$, the variation of which should be sufficiently smooth for the applicability of the quasiclassical approximation. The change in wavelength at a distance equal to its length should be less than the length itself, that is, $\frac{d\lambda}{dr} \cdot \lambda \ll \lambda$ or

$$\frac{d\lambda}{dr} \ll 1.$$

As a rule this condition is satisfied because of the smoothness of the nucleus–nucleus interaction potentials $V(r)$.

The meaning of the semiclassical approximation is extremely simple. In classical mechanics, the experiment shown in Fig. 2.8 implies the flow of particles incident on the scattering center with different impact parameters. The energy and momentum of these particles are fixed, which corresponds to an incident plane wave $e^{i\mathbf{kr}}$ in quantum mechanics. The wavefunction of such a particle satisfies the Schrodinger equation and can be calculated at any point of the three-dimensional space with the help of the relation (3.14) and the numerical solution of equations (3.15). If the number of partial waves is large, finding the wavefunction becomes a laborious task (to find the wavefunction at a distance R in (3.14), it is necessary to carry out the summation up to $l \le kR$). In this case, however, the quasiclassical approximation and the corresponding wavefunction at (r, θ) are applicable. For example, in Fig. 3.6, we can write

$$\psi_{\vec{k}}^{(+)}(r, \theta) \approx A_1(b_1; r, \theta)e^{iS_1(b_1, r, \theta)} + A_2(b_2; r, \theta)e^{iS_2(b_2, r, \theta)}, \qquad (3.21)$$

where $\hbar S_i(b_i; r, \theta) = \hbar \int_{tr_i} \mathbf{k}(\mathbf{r})d\mathbf{r}$ are the classical action functions computed along the trajectories with impact parameters b_1 and b_2 that pass through the point (r, θ) (the wavefunction does not depend on the azimuthal angle φ in scattering by a central field).

The amplitudes of the two waves in Eq. (3.21) are determined by the density of the trajectories, that is, by the continuity equation. If three trajectories pass through a given point (r, θ), then there will be three terms in Eq. (3.21), and so on. If, however,

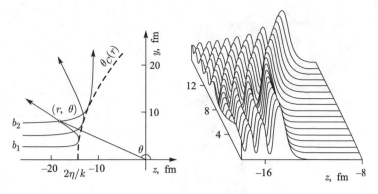

Fig. 3.6 The field of classical trajectories (the dashed line shows the caustic surface separating the classically inaccessible region) and the amplitude of the wavefunction in the scattering of particles by the Coulomb field ($\eta = 57.1$, $k = 7.93\,\text{fm}^{-1}$)

a trajectory passes through the region of the absorbing optical-model potential $W(r)$, then the local wave number $k(r) = \sqrt{\frac{2\mu}{\hbar^2}[E - V(r) - i\,W(r)]}$, and hence, the action function acquires a positive definite imaginary part, which decreases the amplitude of the corresponding wave.

When the particles are scattered by the Coulomb field (see Figs. 3.1b and 3.6) just two trajectories pass through each point of the three-dimensional space (r, θ) in the classically allowed region ($\theta > \theta_C(r)$, or $r > r_C(\theta)$). For Coulomb trajectories, all quantities are known in explicit form, and in the quasiclassical approximation the corresponding wavefunction can be written in a simple analytical form [76]. The amplitude of this function is shown in Fig. 3.6. In the classically allowed region, we have interference of two terms: an incident wave with amplitude $A_1(b_1, r \to \infty) \to 1$ and a reflected wave with amplitude $A_2(b_2, r \to \infty) \to f_C(\theta)/r$, where f_C is the Coulomb scattering amplitude (2.14). In the classically forbidden region, the Coulomb wavefunction is approximated by the Airy function and decreases exponentially as expected in quantum mechanics. Near the caustic surface $\theta_C(r)$, separating the classically allowed and forbidden regions, the amplitude of the wavefunction sharply increases. When the wavelength is decreasing $\lambdabar \to 0$, the amplitude of the Coulomb wavefunction increases as $\lambdabar^{-1/6}$ [51]:

$$\left| \Psi_{\mathbf{k}}^{C(+)}(r, \theta_C(r)) \right| = \sqrt{\pi}\,Ai(0)(2\eta)^{1/6} \approx 0.3550(2\eta)^{1/6},$$

where $Ai(z)$ is the Airy function and η is the Sommerfeld parameter.

Such an increase in the amplitude of a wave with a decrease in its wavelength is called a wave catastrophe. This phenomenon is inherent in all wave processes in which reflective waves appear in the caustic surfaces, for example, reflection of radio waves from the upper layers of the ionosphere, or reflection of acoustic waves from boundaries with different densities. The increase in the amplitude of the wave depends on the nature of the caustic surface. When the two simple caustics cross

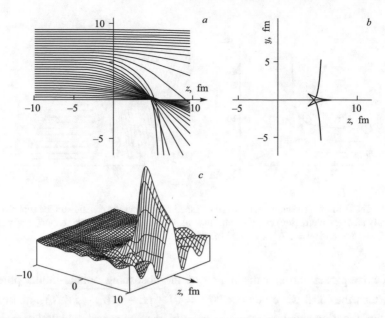

Fig. 3.7 (**a**) The field of classical trajectories, (**b**) the caustic surface, and (**c**) the amplitude of the wavefunction for the scattering of neutrons with an energy of 50 MeV on the ^{40}Ca nucleus

(merge), a surface is created that is called the *caustic beak*. The amplitude of the wave in the region of this beak is approximated by the resulting etalon integral, the Pierce function, that increases as $\lambda^{-1/4}$ with decreasing λ [76]. In the scattering of neutrons that deflected to negative angles in the mean field and focus onto the beam axis in the region behind the nucleus, an even more complex *butterfly* type caustic surface can appear (the so-called A_5 catastrophe), and the wave amplitude grows with decreasing λ as $\lambda^{-1/3}$ [51], see Fig. 3.7.

3.5   Nuclear Rainbow and Diffraction Scattering

For small wavelength (large wave numbers k), when a large contribution to the sum in Eq. (3.19) comes from many partial waves, the scattering amplitude and the differential cross section for elastic scattering can also be calculated in a quasiclassical approximation. In addition to providing simplicity, such calculations, based on the trajectory description of the scattering process, also allow us to understand the character of the angular dependence of the observed cross section. If the scattering process takes place in a central field containing repulsion at large distances and attraction at small ones (corresponding to the scattering of charged nuclei at above-barrier energies, see Fig. 3.1), then the classical deflection function looks as shown in Fig. 3.8. (This figure does not show the region of very small impact parameters,

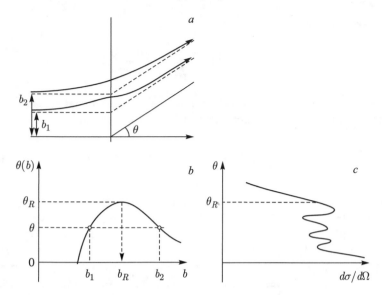

Fig. 3.8 (**a**) Two trajectories leading to scattering at the same angle θ. (**b**) A schematic representation of the deflection function versus the impact parameter for some central field that repels particles at large distances and attracts at small ones. (**c**) Differential elastic scattering cross section corresponding to the deflection function shown in figure (**b**). The angle θ_R is referred to as the rainbow angle

for which the colliding nuclei overlap significantly and are essentially absorbed from the elastic channel). In this case, there are two trajectories with different impact parameters leading to scattering at the same angle (see Fig. 3.8a), and as the angle θ decreases, the larger impact parameter $b_2(\theta)$ approaches the corresponding impact parameter of the Coulomb trajectory $b_C(\theta) = \frac{\eta}{k}\cot(\theta/2)$ (see Sect. 3.1). In addition, the deflection function has a characteristic maximum at θ_R, that is, there are no classical trajectories leading to scattering at angles $\theta > \theta_R$.

In contrast to Fig. 3.2, for each angle $\theta < \theta_R$ there are two *tubes* of trajectories with impact parameters close to $b_1(\theta)$ and $b_2(\theta)$ that will impinge on a distant detector in the solid angle $\Delta\Omega$. The contribution of these two tubes leads to two terms in the classical differential cross section for elastic scattering

$$\frac{d\sigma^{cl}}{d\Omega}(\theta) = \frac{b_1(\theta)}{\sin\theta}\frac{1}{|d\theta(b_1)/db|} + \frac{b_2(\theta)}{\sin\theta}\frac{1}{|d\theta(b_2)/db|} = \sigma_1^{cl}(\theta) + \sigma_2^{cl}(\theta). \quad (3.22)$$

In the optical model of elastic scattering, a particle passing through the absorbing potential $W(r)$ has a certain probability to be removed from the elastic channel (that is, some nuclear reaction will take place). When moving along a trajectory with impact parameter b, the probability of remaining in the elastic channel is determined by the mean free path:

$$P_{els}(b) = \exp\left[-\int_{tr(b)} \frac{ds}{\lambda_{fr}(r)}\right],$$

where $\lambda_{fr} = -\frac{\hbar v}{2W(r)}$. Thus, in the general case, the differential cross section for elastic scattering in classical mechanics has the following form

$$\frac{d\sigma_{cl}}{d\Omega}(\theta) = \sum_i \frac{b_i(\theta)}{\sin\theta} \frac{P_{els}(b_i(\theta))}{|d\theta(b_i)/db|} \equiv \sum_i \sigma_i^{cl}(\theta) P_{els}(b_i(\theta)), \qquad (3.23)$$

where the summation is performed over all trajectories leading to scattering at the angles $\pm\theta\pm 2n\pi$, since experimentally, these scattering angles are indistinguishable.

As the angle θ approaches the angle θ_R, the two terms in Eq. (3.22) converge and become infinitely large because the denominator vanishes. This is nothing but the focusing of trajectories: a large number of trajectories with impact parameters in the interval $b_R \pm \Delta b$ lead to scattering to an angle $\Delta\theta$ infinitesimally close to θ_R ($\Delta b/\Delta\theta \to \infty$, see Fig. 3.8). A sharp increase in the scattering cross section near the angle θ_R is referred to as rainbow scattering, and the angle θ_R is called the rainbow scattering angle, since it is this focusing effect that leads to the well-known natural phenomenon caused by the refraction of light rays in droplets of rainwater. The angle θ_R in this case turns out to be negative and less than $-90°$ (backward scattering). A similar phenomenon is observed in the scattering of nuclear particles (see below).

All quantum effects can be taken into account in the quasiclassical approximation, in which fairly simple expressions for the partial phases and scattering amplitudes are obtained. In particular, the Coulomb phase shifts are given by the formula

$$\sigma_l^{qcl} = \eta \ln\sqrt{(kb)^2 + \eta^2} + kb \cdot \arctan(\eta/kb) - \eta,$$

and the sum of the nuclear and Coulomb phases is calculated using the expression

$$\chi_l = \delta_l + \sigma_l = kb\frac{\pi}{2} - kr_0(b) + \eta \ln 2kr_0 + \int_{r_0(b)}^\infty \left[k(b,r) - k + \frac{\eta}{r}\right] dr, \qquad (3.24)$$

where $k(b,r) = k\sqrt{1 - V(r)/E - b^2/r^2}$ is the local wave number, and the classical turning point $r_0(b)$ (the distance of closest approach) of the trajectory for impact parameter $b = (l + 1/2)/k$ is found from the condition $k(b, r_0) = 0$. In the general case $r_0(b)$ is a complex quantity whose imaginary part arises from the possibility of above-barrier reflection of the wave and because of the imaginary part of the optical-model potential. An important feature of the expression (3.24) for the scattering phase shift is the relation

$$2\frac{\partial \chi_l}{\partial l} = \theta(b) \equiv \pi - 2 \int_{r_0(b)}^{\infty} \frac{kb\,dr}{r^2 k(b,r)}, \tag{3.25}$$

which makes it possible to calculate the scattering amplitude in the semiclassical approximation. The right-hand side of (3.25) is simply the deflection function for impact parameter b in the potential $V(r)$, where $kb = l + 1/2$.

As already noted, for the scattering of heavy particles (or particles with high energy), many partial waves contribute to the scattering cross section and formally the summation over l in Eq. (3.19) can be replaced by an integral. For large values of l and angles in the range $1/l < \theta < \pi - 1/l$, the Legendre polynomials have the form

$$P_l(\theta) \approx \sqrt{\frac{2}{\pi \lambda \sin \theta}} \frac{1}{2} \left[e^{i(\lambda\theta - \pi/4)} + e^{-i(\lambda\theta - \pi/4)} \right], \tag{3.26}$$

(where we have introduced the notation $\lambda = l + 1/2$), and the total scattering amplitude can be written

$$f(\theta) = \frac{1}{2ik} \sum_{l=0}^{\infty} (2l+1) \left(e^{2i\chi_l} - 1 \right) P_l(\cos \theta)$$

$$\approx \frac{1}{2ik\sqrt{\pi \sin \theta}} \sum_{l=0}^{\infty} \sqrt{2\lambda} e^{2i\chi(\lambda)} \left[e^{i(\lambda\theta - \pi/4)} + e^{-i(\lambda\theta - \pi/4)} \right]. \tag{3.27}$$

The term -1 in brackets in the sum over l can be ignored in Eq. (3.27), since it gives zero for all angles except $\theta = 0$: $\sum_l (2l+1) P_l(\cos \theta) = 2\delta(1 - \cos \theta)$.

We now replace the sum over l by the corresponding integral over λ. It is well known that in calculating integrals of strongly oscillating functions, the main contributions come from the stationary points of the phases of these functions. In our case, in the integral over λ, we have two terms with phases $2\chi(\lambda) + \lambda\theta$ and $2\chi(\lambda) - \lambda\theta$. The stationary condition for these phases, $\partial/\partial\lambda [2\chi(\lambda) \pm \lambda\theta] = 0$, determines which partial waves (or impact parameters) give the main contributions to the scattering at an angle θ. Taking into account Eq. (3.25), these conditions seem quite obvious: $\theta(b = \lambda/k) = \pm\theta$, that is, the main contributions to scattering at an angle θ are provided by impact parameters that classically lead to scattering at that angle or at the angle $-\theta$. Near such stationarity points, the corresponding phase shifts can be approximated by the quadratic

$$2\chi(\lambda) \pm \lambda\theta \approx 2\chi(\lambda_i) \pm \lambda_i\theta + \chi''(\lambda_i)(\lambda - \lambda_i)^2,$$

and the quasiclassical scattering amplitude calculated by this stationary phase method is

$$f^{qcl}(\theta) = \sum_i \left(\frac{b_i(\theta)}{\sin\theta} \frac{1}{|d\theta(b_i)/db|} \right)^{1/2} \exp\left[2i\chi(\lambda_i) \pm \lambda_i\theta \pm \pi/2\right]. \tag{3.28}$$

We see that the square of the modulus of each of the terms in Eq. (3.28) coincides with the corresponding classical cross section (the imaginary part of the scattering phase χ_l ensures the absorption). However, unlike the classical cross section Eq. (3.23), the expression Eq. (3.28) takes into account the wave properties of the scattered particles, that is, the total scattering amplitude is a coherent sum of amplitudes and $d\sigma/d\Omega = |f(\theta)|^2$. In particular, in the presence of two trajectories (that is, two impact parameters $b_1(\theta) \neq b_2(\theta)$) leading to scattering at an angle θ, a characteristic interference pattern appears in the angular distribution, see Fig. 3.8c. Depending on the path difference of the waves, that is, on the phase difference in expression Eq. (3.28), the two waves can add in phase or out of phase and thus amplify or suppress one another.

When approaching the rainbow scattering angle, two stationary points approach each other $b_1 \to b_2 \to b_R$ and $2\chi''(b = b_R) = \theta'(b = b_R) = 0$. The total phase $2\chi(\lambda) \pm \lambda\theta$ at $b \approx b_R$ must now be approximated by a cubic expansion, and the differential cross section for elastic scattering can be written in the form [14]

$$\frac{d\sigma^{qcl}}{d\Omega}(\theta \approx \theta_R) = \frac{2\pi b_R}{k\sin\theta} \cdot |C|^{-2/3} \cdot Ai^2(z) \cdot P_{els}(b_R), \tag{3.29}$$

where $Ai(z)$ is the Airy function, the coefficient $C = \frac{1}{2k^2}\frac{d^2\theta}{db^2}\big|_{b=b_R}$ and $z = \pm|C|^{-1/3}(\theta - \theta_R)$ (the plus sign is selected in the case of a maximum, and minus in the case of a minimum of the deflection function at the point θ_R). This cross section decreases exponentially from the dark side of the rainbow ($z > 0$) and oscillates on its light side ($z < 0$), reaching a maximum not at $\theta = \theta_R$, but at $z \approx -1$, see Fig. 3.8c.

The case of elastic scattering of ^{3}He nuclei on ^{14}C at an energy of 24 MeV/nucleon serves as an experimental example of the rainbow scattering of nuclei [20] (Fig. 3.9).

In the case of elastic scattering of relatively light ions with high energy, a typical diffraction pattern is observed, caused by the strong absorption of all trajectories with small impact parameters (scattering on an opaque disk). A typical picture of the diffraction scattering of ^{20}Ne nuclei at an energy of 390 MeV on a carbon target is shown in Fig. 3.10a.

An approximate expression for the scattering cross section can be obtained if it is assumed that all the partial waves with $l \leq l_0(b \leq R_W \approx R_1 + R_2)$ are completely *absorbed* (that is, lead to some other reaction channels) giving $S_l(l \leq l_0) = 0$, and the partial waves with $l > l_0$ are not distorted by the scattering potential (that is, for impact parameters $b > l_0/k \approx R_1 + R_2$ the nuclei do not interact) and

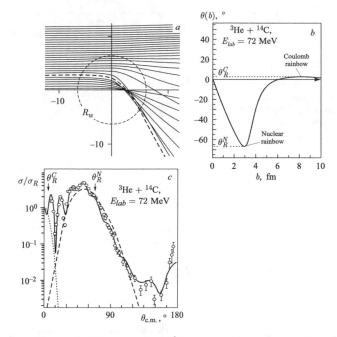

Fig. 3.9 (**a**) The trajectory field for scattering of ^{3}He nuclei with an energy of 24 MeV/nucleon by ^{14}C. The dashed curve corresponds to the impact parameter of the nuclear rainbow b_R^N. The circle shows the radius of the absorbing potential. (**b**) The classical deflection function. The angles of the Coulomb ($\theta_R^C \approx +2°$) and nuclear ($\theta_R^N \approx -67°$) rainbows are shown. (**c**) Differential cross section for elastic scattering of ^{3}He nuclei on ^{14}C at the energy of 24 MeV/nucleon. The experimental data from [20] are shown by dots, the solid curve is obtained in the OM with the parameters indicated in the text, and the dashed and dotted curves show the contributions of the nuclear and Coulomb rainbows calculated by the formula (3.29)

$S_l(l > l_0) = 1$. This is not far from the truth; if we look at the sharp dependence on l of the precisely calculated partial-wave S-matrix (Fig. 3.10b). In this case $f_l = 2\pi(S_l - 1)/ik = 0$ for $l > l_0$, and the nuclear scattering amplitude (2.13) takes the form

$$f_N(\theta) \approx -\frac{1}{2ik} \sum_{l=0}^{l_0} (2l + 1) P_l(\cos\theta), \qquad (3.30)$$

and the Coulomb scattering can be neglected, since, for the light particles and high energies the Coulomb parameter η is sufficiently small so that the trajectories for $b > R_W$ are practically undeflected. Using the property of the Legendre polynomials

$$(2l + 1) P_l(x) = \frac{d}{dx} [P_{l+1}(x) - P_{l-1}(x)]$$

Fig. 3.10 (a) Differential cross section for elastic scattering of ^{20}Ne nuclei with the energy of $E_{lab} = 390\,\text{MeV}$ on a ^{12}C target [14]. The solid curve shows an optical-model calculation. (b) Partial elements of the elastic scattering S-matrix ^{20}Ne + ^{12}C at $E_{lab} = 390\,\text{MeV}$

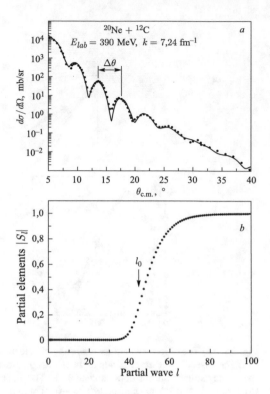

for $l \geq 1$ and $P_0(x) = 1$, one can reduce the expression (3.30) to the very simple form

$$f(\theta) = \frac{i}{2k}\frac{d}{dx}\left[P_{l_0+1}(x) + P_{l_0}(x)\right] \approx \frac{i}{2k}\frac{d}{dx}P_{l_0}(x).$$

For values $l \gg 1$, the Legendre polynomials behave as the Bessel function of zero order, $P_l(\cos\theta) \approx J_0(l\theta)$. Since $dJ_0(x)/dx = J_1(x)$, we then have $f(\theta) \approx i(l_0/k)J_1(l_0\theta)/\theta$ and

$$\frac{d\sigma}{d\Omega} = |f(\theta)|^2 \approx R_W^2\left[\frac{J_1(kR_W\theta)}{\theta}\right]^2, \tag{3.31}$$

where $R_W = l_0/k$ is the radius of the absorbing disk on which the plane wave is incident. Expression (3.31) is identical to the optical Fraunhofer diffraction on a black disk. For large arguments, the square of the Bessel function $J_1^2(x)$ is a periodic function with period π, that is, in the angular distribution of elastic scattering, a decrease in the amplitude maxima should be observed with a period $\Delta\theta = \pi/kR_W$, as we see in Fig. 3.10a. Taking the value $\Delta\theta \approx 4°$ from the figure, we obtain (for a wave number $k = 7.24\,\text{fm}^{-1}$) the value $R_W \approx 6\,\text{fm}$ for the radius of the absorbing disk, which looks reasonable.

3.6 Elastic Scattering of Heavy Ions

Elastic scattering involving heavy ions is determined by two main factors: strong Coulomb repulsion and a high probability of loss (absorption) from the elastic channel for all trajectories (partial waves) for which the colliding nuclei come into contact. In this case, the use of classical trajectories and of the classical deflection function further simplifies our understanding and interpretation of the experimental data.

For large values of Z_1 and/or Z_2, the repulsive Coulomb forces lead to significant deviations from linear motion even at large distances. The panel (a) of Fig. 3.11 shows the field of trajectories of the relative motion for ^{16}O and ^{88}Sr nuclei at an energy $E_{c.m.} = 47.4$ MeV (Coulomb parameter $\eta = 25.6$). At large impact parameters, the nuclei move along Coulomb trajectories. The deflection function (shown in the panel (b) of the figure) coincides with the expression (3.3) for scattering in the Coulomb field and, up to some angles, the elastic scattering cross section coincides with the Rutherford cross section Eq. (3.5) (the ratio of the differential cross section to the Rutherford cross section $d\sigma/d\sigma_R$ is shown on the panel (c) of Fig. 3.11). With a decrease in the impact parameter, the colliding nuclei approach each other to distances at which the attractive nuclear forces start to act.

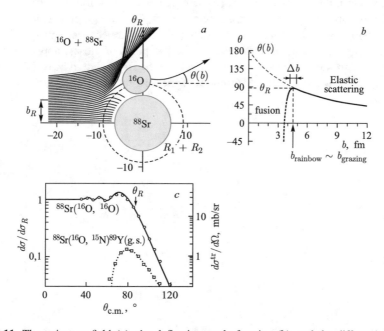

Fig. 3.11 The trajectory field (**a**), the deflection angle function (**b**), and the differential cross section for elastic scattering of ^{16}O oxygen nuclei with energy $E_{lab} = 56$ MeV on a strontium target ^{88}Sr (experimental data were taken from [4]) (**c**). The figure (**c**) also shows the cross section of the proton transfer process (the scale is shown on the right, see Chap. 3)

The deflection function reaches a maximum (for some $b = b_R$) and then begins to decrease sharply. The dashed line describing this function in Fig. 3.11 reflects the fact that when $b < b_R$ there is no elastic scattering: the surfaces of the nuclei overlap and some nuclear reaction will occur (in this case, mainly the fusion of the two nuclei). As a result, the elastic scattering cross section decreases sharply for $\theta > \theta_R$.

The presence of an extremum in the deflection function leads to the focusing of trajectories and is called rainbow scattering (see the previous paragraph). In contrast to the case of elastic scattering of light ^{3}He nuclei on ^{14}C considered above (see Fig. 3.9), in the present reaction the focusing of trajectories occurs at large positive angles (due to strong Coulomb repulsion), so this phenomenon is called the *Coulomb* rainbow. On the bright side of the rainbow, $\theta < \theta_R$, a certain increase in the differential cross section of elastic scattering and a weakly pronounced interference phenomenon, qualitatively described by the Airy function in Eq. (3.29), are observed. A sharp increase in the cross section for the elastic scattering of heavy ions at $\theta \sim \theta_R$ does not occur because trajectories with impact parameters $b \sim b_R$ already fall within the range of action of nuclear forces and, with a high probability, are lost from the elastic channel, into some reaction channel.

In describing collisions of heavy ions, it is customary to introduce an impact parameter corresponding to a grazing collision b_{gr}, such that for $b > b_{gr}$, mainly elastic scattering takes place, and for $b < b_{gr}$, nuclear reactions occur with a high probability. Obviously the two impact parameters, b_R and b_{gr}, practically coincide, since in both cases the main condition is the passing of trajectories with $b \sim b_R, b_{gr}$ into the range of the nuclear forces, that is, the point of closest approach of the trajectory with such an impact parameter $r_0(b \sim b_R, b_{gr})$ should be of the order of $R_1 + R_2 + 2a$, where a is the diffuseness of the nuclear surface.

Chapter 4
Quasi-Elastic Scattering of Heavy Ions and Few-Nucleon Transfer Reactions

For low-energy nucleus-nucleus collisions, the process of complete fusion (for light and medium-sized nuclei, see Chap. 6) and binary processes with the formation of two final nuclei in the exit channel usually dominate. (Only for weakly bound light nuclei the cross section of the breakup channel commensurate with those for the binary channels). Suppose that when two nuclei, a and A, collide, two other nuclei, b and B, are formed: $a + A \rightarrow b + B$. In the experiment, as a rule, charge, mass, angular, and energy distributions of the products of such a reaction are measured to study the dynamics of the collision. In some cases (discussed in this chapter), it is possible to identify a channel with a given excited state of the final nucleus B, $\Phi_{\nu_f}(B)$. In that case, the charge, mass and energy of nucleus b are strictly specified (determined by the conservation laws) and only its angular distribution needs to be measured. From this, information on the structure of the studied nucleus B can be extracted. The differential cross section for such a process is determined by the corresponding transition amplitude and has the form

$$\frac{d\sigma_{fi}}{d\Omega}(\theta) = \frac{1}{(2\pi)^2} \frac{\mu_i \mu_f}{\hbar^4} \frac{k_f}{k_i} \left| T_{fi}(\vec{k}_f, \vec{k}_i) \right|^2. \tag{4.1}$$

Here, $\vec{k}_i$ and $\vec{k}_f$ are the momenta of the relative motion of the incoming and outgoing nuclei, θ is the angle between the vectors $\vec{k}_i$ and $\vec{k}_f$; μ_i and μ_f are the reduced masses in the entrance and exit channels, and the subscripts i, f include the quantum numbers of the initial and final states of the nuclei. The transition amplitude T_{fi} is determined by the standard expression

$$T_{fi}(\vec{k}_f, \vec{k}_i) = \left\langle e^{i\vec{k}_f \vec{R}_f} \Phi_{\nu_f'}(b) \Phi_{\nu_f}(B) \left| V_{bB} \right| \Psi_{\nu_i \vec{k}_i}^{(+)} \right\rangle. \tag{4.2}$$

The internal wavefunctions of the nuclei $\Phi_\nu(c)$ satisfy the Schrodinger equations for bound states: $H_c \Phi_\nu(c) = \varepsilon_\nu^c \Phi_\nu(c)$, where $c = a, A, b, B$. The function $\Psi_{\nu, \vec{k}}^{(+)}$

© Springer Nature Switzerland AG 2019
V. Zagrebaev, *Heavy Ion Reactions at Low Energies*, Lecture Notes in Physics 963,
https://doi.org/10.1007/978-3-030-27217-3_4

is the total wavefunction of the system with a boundary condition in the form of an incident wave in the entrance channel and outgoing spherical waves in all the exit channels. This function satisfies the Schrodinger equation with the total Hamiltonian of the system

$$H = -\frac{\hbar^2}{2\mu_i}\nabla^2_{\vec{R}_i} + H_a + H_A + V_{aA} = -\frac{\hbar^2}{2\mu_f}\nabla^2_{\vec{R}_f} + H_b + H_B + V_{bB}, \qquad (4.3)$$

in which $\vec{R}_i = \vec{r}_a - \vec{r}_A$, $\vec{R}_f = \vec{r}_b - \vec{r}_B$, and V_{aA} and V_{bB} are the interactions between the nuclei, that depend not only on the separations, R_i and R_f, but also on the internal variables (that is, on all the coordinates of the nucleons making up these nuclei). The angular brackets in Eq. (4.2) mean integration over all these variables. If unpolarized beams and targets are used in the experiment, and polarization of the final nuclei is not measured, then, strictly speaking, the cross section (4.1) must be averaged over all projections of the spins of the initial nuclei and summed over the projections of the spins of the final nuclei.

Precise calculation of the transition amplitude Eq. (4.2) is impossible, not only because of technical problems (too many variables), but also because of our incomplete knowledge of the exact form of nucleus–nucleus interactions and of the complex structure of the nuclei themselves. However, in many nuclear reactions the situation is much simpler. Such cases are just the direct processes of inelastic excitation and of direct transfer reactions.

4.1   Direct Process of Light-Particle Transfer

The term *direct nuclear reactions* usually refers to the processes in which, in addition to the relative motion of the colliding nuclei, only one or a small number of degrees of freedom are involved. In the collision of light particles, these processes occur with high probability and, very importantly, are fairly easy to separate experimentally. In reactions with heavy ions, direct processes are rather an exception and can occur only in a narrow range of impact parameters that correspond to grazing collisions. With the help of direct nuclear reactions, it is possible to selectively study the properties of specific states of atomic nuclei, for example, their rotational or vibrational excitations, single-nucleon, or cluster states. The description of direct nuclear reactions, as a rule, is made in the framework of microscopic models, which make it possible to extract the desired information more reliably.

As an example, let us consider the process of transfer of a particle x from an incident nucleus a, having a momentum $\vec{k}_i$ in the input channel, to the nucleus b with the formation of a final nucleus B and emergence of a fragment b with a momentum $\vec{k}_f$ in the output channel. A proton or a neutron can act as a particle x. In this case we study single-particle states in the nuclei a and B. Particle x may also be treated

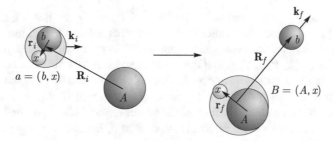

Fig. 4.1 Schematic representation of the direct reaction of transfer of the particle x (a nucleon or light cluster) from the incident nucleus a to the target nucleus A

as some cluster, for example, an alpha particle, if we are interested in the cluster states of these nuclei. If the transfer process occurs without excitation of the internal degrees of freedom of the nuclei a and A, then this is a direct reaction (Fig. 4.1). Such a process is nevertheless often accompanied by an intermediate excitation of the colliding nuclei. In this case, the reaction has a multistage nature and for its description it is necessary to take into account the channel coupling (see below).

The cross section for the transfer of fragment x, $A(a, b)B$, is determined by the expression (4.1), and the corresponding transition amplitude now has a simpler form

$$T_{fi}(\vec{k}_f, \vec{k}_i) = \left\langle e^{i\vec{k}_f \vec{R}_f} \varphi_B^{v_f}(\vec{r}_f) \, | V_{xb} + V_{bA} | \, \Psi_{v_i \vec{k}_i}^{(+)}(\vec{r}_i, \vec{R}_i) \right\rangle. \qquad (4.4)$$

Here $\varphi_B^{v_f}(\vec{r}_f)$ is the wavefunction of the $(A + x)$ bound state with a set of quantum numbers $v_f = n_f, l_f, j_f$ forming in the exit channel. The function $\Psi_{v_i \vec{k}_i}^{(+)}(\vec{r}_i, \vec{R}_i)$ is the total wavefunction of the system with boundary conditions corresponding to an incident wave with momentum $\vec{k}_i$ in the entrance channel, and including the bound state of the particles x and b with quantum numbers $v_i = n_i, l_i, j_i$ that is described by the wavefunction $\varphi_a^{v_i}(\vec{r}_i)$. For simplicity, the magnetic quantum numbers discussed above are omitted here.

The interaction $\Delta V = V_{xb} + V_{bA}$ responsible for the transition is a part of the total interaction of all particles $V_{xb} + V_{xA} + V_{bA}$, minus the interaction V_{xA} already taken into account in the output channel. The angular brackets in the expression (4.4) mean integration over all 6 coordinates (obviously, the variables $\vec{r}_f$, $\vec{R}_f$ are functions depending on $\vec{r}_i$, $\vec{R}_i$).

4.2 Distorted-Wave Description of Direct Reactions

The total wavefunction $\Psi_{v_i, \vec{k}_i}^{(+)}(\vec{r}_i, \vec{R}_i)$ in the transition amplitude (4.4) is still quite difficult to calculate. Its simplest (Born) approximation is the product of an incident plane wave in the input channel with the wavefunction of the bound state of the particles x and b in the incident projectile—$\varphi_a^{v_i}(\vec{r}_i)e^{i\vec{k}_i\vec{R}_i}$. In this approximation, the interactions between the nuclei in the input channel and the exiting fragments in the output channel are completely ignored. This approximation is justified for high energies ($E \gg V_{aA}, V_{bB}$) and light particles. Even for collisions between ions of medium mass, a strong deviation of the relative motion from linear motion is observed (see, for example, Fig. 3.11) as well as a high probability of absorption (leaving the elastic channel) for low partial waves (small impact parameters).

It is possible to take these effects into account if, instead of using plane waves in the input and output channels, one uses the so-called distorted waves, $\psi_{\vec{k}_i}^{(+)}(\vec{R}_i)$ and $\psi_{\vec{k}_f}^{(-)}(\vec{R}_f)$, that describe the elastic scattering in the input and output channels, by taking into account the corresponding optical potentials, $U_{aA}(R_i)$ and $U_{bB}(R_f)$. This approximation is called the distorted wave born approximation (DWBA) and the transition amplitude takes the form

$$T_{fi}^{DWBA}(\vec{k}_f, \vec{k}_i)$$

$$= \left\langle \psi_{\vec{k}_f}^{(-)}(\vec{R}_f)\varphi_B^{v_f}(\vec{r}_f) \,|V_{xb} + V_{bA} - U_{bB}|\, \varphi_a^{v_i}(\vec{r}_i)\psi_{\vec{k}_i}^{(+)}(\vec{R}_i) \right\rangle. \qquad (4.5)$$

The symbol $(-)$ in the wavefunction of the output channel $\psi_{\vec{k}_f}^{(-)}$ refers to the boundary conditions (formally, an ingoing spherical wave). However, since $\psi_{\vec{k}}^{*(-)} = \psi_{-\vec{k}}^{(+)}$ (see textbooks on scattering theory), the wavefunction $\psi_{\vec{k}_f}^{*(-)}(\vec{R}_f)$ entering into Eq. (4.5) is calculated in exactly the same way as $\psi_{\vec{k}_i}^{(+)}(\vec{R}_i)$ using the partial-wave expansion (3.14) and the solution of the radial Schrodinger equation for partial wavefunctions. Thus when $\vec{k}$ is replaced by $-\vec{k}$, an angle θ between the vectors $\vec{r}$ and $\vec{k}$ changes to $\pi - \theta$, and $P_l(-\cos\theta) = (-1)^l P_l(\cos\theta)$, then in the expansion in terms of partial waves of the function $\psi_{\vec{k}_f}^{*(-)}(\vec{R}_f)$, the factor i^l must simply be replaced by i^{-l} in Eq. (3.14).

Further simplification of the transition amplitude (4.5) usually relates to the assumption of the similarity between the interactions V_{bA} and U_{bB} (in the case of transfer of a single-nucleon, $x = n$ or p, it is really true), that results from neglecting the contribution of $V_{bA} - U_{bB}$ to the particle-transfer process. After the expansion in terms of the partial waves of the functions $\psi_{\vec{k}_i}^{(+)}(\vec{R}_i)$ and $\psi_{\vec{k}_f}^{*(-)}(\vec{R}_f)$, the radial integrals are computed numerically. The integration over angular variables is carried out in analytic form, taking into account the orthogonality and completeness of the spherical harmonics. One of the most popular computational codes that calculates cross sections for the direct process of light-particle transfer within the framework of the distorted wave method is the DWUCK program [40]. Calculations using the interactive version of this program can be carried out directly on the Internet [http://nrv.jinr.ru/nrv] by any remote user.

4.3   Single-Particle States and Cluster States, Spectroscopic Factors

One of the main advantages of direct transfer processes is the possibility of studying single-particle and cluster properties of atomic nuclei. The cross section of the transfer process is determined not only by the distorted waves but also by the properties of the bound states of the initial and final nucleus $\varphi_a^{v_i}(\vec{r}_i) = f_{n_i l_i}(r_i) \cdot Y_{l_i m_i}(\Omega_i)$ and $\varphi_B^{v_f}(\vec{r}_f) = f_{n_f l_f}(r_f) \cdot Y_{l_f m_f}(\Omega_f)$. In single-step reactions with light ions, the angular distribution of the final particles b strongly depends on the value of the transferred orbital momentum $\vec{\lambda} = \vec{l}_f - \vec{l}_i$. If the situation with spins is fairly simple (even-even nuclei, zero angular momentum of the initial state, etc.), then from the form of the angular distribution of the differential cross Sect. 4.1, the angular momentum (and sometimes other quantum numbers) of the state v_f being populated can be extracted.

This is particularly easy to understand from the example of (d, p) reactions. First, in this case, taking into account the short-range of the nucleon–nucleon forces, one can use the so-called zero-range approximation:

$$V_{xb}(\vec{r})\varphi_a^{v_i}(\vec{r}) \equiv V_{pn}(\vec{r})\varphi_d(\vec{r}) = D_0\delta^3(\vec{r}),$$

that leads to the reduction of the 6-dimensional integral to a 3-dimensional one. Secondly, one can account for the peripheral character of the single-step reaction by assuming that at small distances the distorted waves $\psi_{\vec{k}_d}^{(+)}(\vec{R})$ and $\psi_{\vec{k}_p}^{*(-)}(\vec{R})$ are absorbed with a high probability, and at large distances $(r > R_W)$ they can be replaced by plane waves so that $\psi_{\vec{k}_p}^{*(-)}(\vec{R}) \cdot \psi_{\vec{k}_d}^{(+)}(\vec{R}) \approx e^{-i(\vec{k}_p - \vec{k}_d)\vec{R}}$. In this case, the transition amplitude is reduced to a simple expression

$$T_{fi}^{DWBA}(\vec{k}_f, \vec{k}_i) \approx const \int_{r > R_W} f_{n_n l_n}(r) \cdot Y_{l_n m_n}(\Omega_r)e^{-i\vec{q}\vec{r}}d^3\vec{r},$$

where $\vec{q}(\theta) = \vec{k}_p - \vec{k}_d$ is the transferred momentum (the scattering angle θ is the angle between the vectors $\vec{k}_p$ and $\vec{k}_d$, that is, the angle of proton emission), and $f_{n_n l_n}(r) \sim e^{-\gamma_n r}$ is the radial wavefunction for the bound state of the neutron in the nucleus B ($\gamma_n = \sqrt{2m_n\varepsilon_n/\hbar^2}$, where ε_n is the neutron binding energy). Let us now expand the plane wave $e^{-i\vec{q}\vec{r}}$ over the partial waves as in Eq. (3.14). In this case the spherical Bessel functions $j_l(qr)$ occur instead of the radial functions $\psi_l(r)$. Since the spherical harmonics are orthogonal, $\int Y_{l'm'}^*(\Omega)Y_{lm}^*(\Omega)d\Omega = \delta_{ll'}\delta_{mm'}$, only the single term with $l = l_n$ of the sum over l survives. The remaining radial integral, because of the rapid decay of the function $f_{n_n l_n}(r)$, is approximately equal to the Bessel function at the point $r = R_W$ and the neutron transfer cross section becomes[16]

$$\frac{d\sigma}{d\Omega}(\theta) \sim \left| j_{l_n}(q R_W) \right|^2.$$

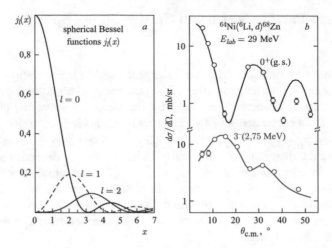

Fig. 4.2 (**a**) Spherical Bessel functions $j_l(x)$ for $l = 0, 1, 2$. (**b**) The angular distribution of deuterons when populating the levels 0^+ (g.s) and 3^- (2.75 MeV) of the ^{68}Zn nucleus in the ^{64}Ni(^{6}Li,d)^{68}Zn transfer reaction at a beam energy of 29 MeV. The solid curves show the calculation of cross sections in the framework of the distorted wave method (4.5) with a *finite range* (nrv.jinr.ru/nrv) and with standard values of the optical potential parameters. The experimental data were taken from [8]

The dependence of the spherical Bessel functions on their argument is shown in Fig. 4.2a for several values of the orbital angular momentum l. It is clear that the character of the angular distribution strongly depends on the orbital angular momentum of the state into which the neutron is transferred, and this orbital angular momentum can be determined by fitting the theoretical angular distribution to the experimental one.

However, for heavier particles, a direct calculation of the amplitude of the transfer process Eq. (4.5) still makes it possible to correctly determine the quantum characteristics of the final nuclei populated. Figure 4.2b shows the angular distributions of deuterons in the transfer of an alpha particle from ^{6}Li on the nucleus ^{64}Ni to form a ^{68}Zn nucleus in its ground state and in a 3^- state with excitation energy of 2.75 MeV.

In addition to the possibility of determining the quantum numbers of the populated states of a final nucleus, direct transfer reactions also make it possible to draw definite conclusions about the so-called spectroscopic factors of these states. It is clear that the real physical state of a final multinucleon nucleus B, $\Phi_\nu(B)$ has a more complex structure than the simple bound state of the fragment x and the core A. However, with some probability, these states are similar to each other. The degree of this similarity (that is, the degree of clusterization of the B nucleus in a $\Phi_\nu(B)$ state to the simple relative motion of the fragments x and A) is simply determined by overlapping the corresponding wavefunctions and is called the spectroscopic factor:

$$S_l^\nu(B \to A + x) = \left| \langle \Phi_{g.s.}(A)\Phi_{g.s.}(x)\varphi_{nl}(\vec{r}_A - \vec{r}_x) | \Phi_\nu(B) \rangle \right|^2.$$

Thus, the differential cross section of the transfer reaction calculated using the distorted wave method must, in the general case, exceed the experimental one:

$$\frac{d\sigma_{fi}^{\exp}}{d\Omega}(\theta) = S^{\nu_f}(B)S^{\nu_i}(a)\frac{d\sigma_{fi}^{DWBA}}{d\Omega}(\theta),$$

see Fig. 4.2b. Using this relation (recorded here in a rather simplified form for spinless particles x and A) in the analysis of experimental data, it is possible to extract the spectroscopic factors of the ground and excited states of atomic nuclei and compare them with those calculated, for example, within the framework of the generalized shell model. This is particularly easy for direct (d, p) reactions. Interest in these reactions has recently increased significantly due to the opportunity to obtain and study the properties of radioactive (proton- and neutron-deficient) nuclei, whose shell properties (including magic numbers) can differ significantly from the properties of nuclei near the stability line.

In principle, the population of some states of the nucleus B can be multistep processes, for example, the intermediate excitation of the colliding nuclei $a + A \rightarrow a' + A_\alpha^* \rightarrow b + B_{\nu_f}$ or the intermediate transfer of nucleons: $a + A \rightarrow c + D_n \rightarrow b + B_{\nu_f}$. In the first case, the problem is solved quite simply: the wavefunction of the entrance channel is decomposed over the orthonormalized basis of the excited states $\Phi_\alpha(A)$ (with $\Phi_{\alpha=0}(A) \equiv \Phi_{g.s.}(A)$, $\langle\Phi_{\alpha'}|\Phi_\alpha\rangle = \delta_{\alpha'\alpha}$), and the distorted wave $\psi_{\alpha=0,\vec{k}_i}^{(+)}(\vec{R}_i)$ (4.5) is replaced by the sum $\sum_\alpha \Phi_\alpha(A)\psi_{\alpha,\vec{k}_i}^{(+)}(\vec{R}_i)$. Usually only a few intermediate excited states are taken into account, and the channel wavefunctions $\psi_{\alpha,\vec{k}_i}^{(+)}$ are found rather simply by solving the relevant coupled Schrodinger equations. For processes with intermediate transfer of nucleons, the situation becomes considerably more complicated due to the non-orthogonality of the states $\Phi_\alpha(A)$ and $\Phi_n(D)$ which are eigenfunctions of different Hamiltonians. A well-known computer code that takes into account the channel coupling in the description of the transmission and inelastic scattering processes is the FRESCO code [67], the use of which, however, requires a sufficiently high qualification and preliminary study.

4.4 Inelastic Excitation of Vibrational and Rotational States

The direct process of inelastic excitation of collective nuclear states looks even simpler. The lower excited states of the nuclei are, as a rule, collective modes of motion: corresponding either to the oscillations (vibrations) of the nuclear surface of the spherical nuclei or to the rotation of statically deformed nuclei. In the first case, the deformation value β_λ is considered as the dynamic variable characterizing the deviation of the nuclear surface from the spherical shape, see Eq. (1.5) and Fig. 1.4. In the second case, it is the rotation angle of the axis of symmetry of the deformed nucleus with respect to some other axis that passes through the nuclear center (see

Fig. 2.4). The excitation of the collective states most likely to be excited (while retaining the same system $a + A \rightarrow a' + A_\alpha^*$ in the entrance and exit channels) is described by the Hamiltonian

$$H = -\frac{\hbar^2}{2\mu_i}\nabla_{\mathbf{r}}^2 + H_A(\xi) + V_{aA}(r, \xi). \qquad (4.6)$$

Here ξ are internal variables describing the excitation of the nucleus A (dynamic deformations and/or angles of rotation). The Hamiltonian of the deformed and rotating nucleus is written in the following form (assuming harmonic oscillations of the surface)

$$H_A = \frac{\hbar^2 \hat{I}^2}{2\mathcal{J}} + \sum_{\lambda \geq 2}\left(-\frac{1}{2d_\lambda}\frac{\partial^2}{\partial s_\lambda^2} + \frac{1}{2}c_\lambda s_\lambda^2\right). \qquad (4.7)$$

Here $\mathcal{J}$ is the moment of inertia of the nucleus A, which determines the kinetic energy of rotation, $s_\lambda = \sqrt{\frac{2\lambda+1}{4\pi}}R_0\beta_\lambda$ is the absolute value of the deformation of multipolarity λ (that is, the maximum deviation of the nuclear radius along the symmetry axis),

$$c_\lambda = C_\lambda(\frac{2\lambda+1}{4\pi}R_0^2)^{-1} = \frac{\hbar\omega_\lambda}{2\langle s_\lambda^0\rangle^2}$$

is the rigidity of the deformable surface, $\langle s_\lambda^0\rangle = \frac{R_0}{\sqrt{4\pi}}\langle\beta_\lambda^0\rangle$ is the amplitude of the zero-point motion, and the mass parameter d_λ is determined from the relation $\hbar\omega_\lambda = \sqrt{\frac{c_\lambda}{d_\lambda}}$ in the liquid-drop model

$$d_\lambda^{LD} = D_\lambda^{LD}\cdot(\frac{2\lambda+1}{4\pi}R_0^2)^{-1} = \frac{3}{\lambda(2\lambda+1)}Am_N.$$

The eigenfunctions of the Hamiltonian (3.7) and its eigenvalues are easily calculated: $H_A\varphi_A^\nu(\xi) = \varepsilon_\nu\varphi_A^\nu(\xi)$. In the case of rotations $\varepsilon_I = \frac{\hbar^2}{2\mathcal{J}}I(I+1)$, and the eigenfunctions are $\varphi_A^{IM}(\theta, \phi) \sim Y_{IM}(\theta, \phi)$; in the case of harmonic oscillations of the nuclear surfaces $\varepsilon_n^\lambda = \hbar\omega_\lambda(n + 3/2)$ and $\varphi_A^n(\xi)$ are expressed in terms of the Hermite polynomials. The transition amplitude (4.2) for the inelastic excitation of the nucleus A has the form

$$T_{\nu 0}(\vec{k}_f, \vec{k}_i) = \left\langle e^{i\vec{k}_f\vec{r}}\varphi_A^\nu(\xi)\,|V_{aA}(r, \xi)|\,\Psi_{0\vec{k}_i}^{(+)}(\vec{r}, \xi)\right\rangle, \qquad (4.8)$$

where $\Psi_{0\vec{k}_i}^{(+)}(\vec{r}, \xi)$ is the total wavefunction of the system that satisfies the Schrodinger equation with the Hamiltonian (4.6), and the nuclear interaction $V_{aA}(r, \xi)$ is the sum of the Coulomb and nuclear potentials (1.9). In the case of direct (single-stage) excitation of the nucleus A, the total wavefunction can be replaced by a simple expression $\Psi^{(+)}_{0\vec{k}_i}(\vec{r}, \xi) \rightarrow \varphi^{\nu=0}_A(\xi) \cdot \psi^{(+)}_{\vec{k}_i}(\vec{r})$, where $\psi^{(+)}_{\vec{k}_i}(\vec{r})$ is the wavefunction describing the elastic scattering. If now the plane wave in the final channel in (4.8) is replaced by an outgoing wave $\psi^{(-)}_{\vec{k}_f}(\vec{r})$ (the corresponding distorting potential does not enter the amplitude of the transition, since $\left\langle \varphi^{\nu\neq0}_A(\xi)|U_{aA}(r)|\varphi^0_A(\xi) \right\rangle = 0$), so we obtain a simple expression for the transition amplitude in the distorted wave method

$$T^{DWBA}_{\nu 0}(\vec{k}_f, \vec{k}_i)$$

$$= \left\langle \psi^{(-)}_{\vec{k}_f}(\vec{r})\varphi^\nu_A(\xi) \left| V^C_{aA}(r, \xi) + V^N_{aA}(r, \xi) \right| \varphi^0_A(\xi)\psi^{(+)}_{\vec{k}_i}(\vec{r}) \right\rangle. \tag{4.9}$$

Here the Coulomb interaction is determined by the expression (1.10), and the nuclear interaction is chosen in the form of a Woods–Saxon potential (1.11) or the proximity potential (1.12). It is the dependence of the Coulomb and nuclear interaction potential on the variable ξ that leads to the possibility of inelastic transitions from the state $\varphi^\nu_A(\xi)$ to the state $\varphi^\mu_A(\xi)$, that is, to the appearance of dynamic deformations of nuclei and/or their rotation. Coulomb and nuclear excitation occurs simultaneously and develops coherently, which leads to interference patterns in the angular distributions of the detected particles.

As an example, Fig. 4.3 shows the angular distribution of carbon nuclei when they are scattered from a ^{144}Nd target following the excitation of a 3^- state with energy 1.51 MeV in the ^{144}Nd nucleus. Excitation of the nucleus due to short-range nuclear forces has a pronounced peripheral (quasi-elastic) character with a maximum in the region of the angle of grazing collisions (see the next section). However, Coulomb excitation occurs on average at large distances due to the long-range terms in formula (1.10). The first term in this formula, the Coulomb interaction of spherical nuclei $\frac{Z_1 Z_2 e^2}{r}$, does not depend on ξ and cannot lead to nuclear excitation. The largest contribution is made by the second term, which slowly decreases at large distances (like $r^{-\lambda}$) and leads to the possibility of inelastic excitation of nuclei even for large impact parameters that lead to scattering at small angles. As a result, the Coulomb excitation has a wider angular distribution shifted toward smaller angles (dashed line in Fig. 4.3). The sum of the Coulomb and nuclear excitation leads to an interference in the cross section of inelastically scattered particles and agrees perfectly with the experimental data.

For the sake of simplicity, the excitation of only one nucleus A was considered above. In the collision of heavy ions, both nuclei can be excited. In this case, we simply add the same Hamiltonian H_a to the Hamiltonian H_A and replace the functions $\varphi^\nu_A(\xi)$ by the function $\varphi^\nu_A(\xi_A)\varphi^\mu_a(\xi_a)$. In the case of excitation of collective nuclear states, it is fairly simple to take into account the strong coupling of the channels, that is, the processes of multiple excitation and de-excitation of nuclei. To do this, it is sufficient to expand the total wavefunction over the excited states

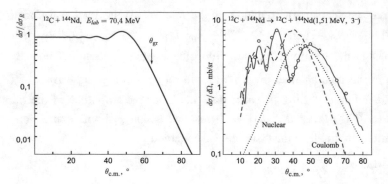

Fig. 4.3 Elastic and inelastic scattering (excitation of the 3^- state at energy 1.51 MeV in the nucleus ^{144}Nd) of carbon nuclei with energy of 70.4 MeV on the target ^{144}Nd. The dashed and dotted curves show contributions to the excitation of the Coulomb and nuclear fields, respectively. The experimental data were taken from [31], and the calculations are performed on the site nrv. jinr.ru/nrv

$\Psi_{0\vec{k}_i}^{(+)}(\vec{r}, \xi_A, \xi_a) = \sum_{v,\mu} \psi_{v,\mu,\vec{k}_i}^{(+)}(\vec{r})\varphi_A^v(\xi_A)\varphi_a^\mu(\xi_a)$ and solve the system of coupled Schrodinger equations for the channel wavefunctions $\psi_{v,\mu,\vec{k}_i}^{(+)}(\vec{r})$ by calculating the matrix elements

$$V_{v'\mu',v\mu}(r) = \left\langle \varphi_A^{v'}\varphi_a^{\mu'} | V_{aA}(r, \xi_A, \xi_a) | \varphi_A^v\varphi_a^\mu \right\rangle.$$

Such a procedure must necessarily be done for rotational excitations with large values of the angular momentum I or vibrational states with phonon numbers $n > 1$. Coupling of the relative motion to the excitation of nuclear collective states also plays a key role in the process of sub-barrier fusion of atomic nuclei (see below).

4.5   Quasi-Elastic Scattering of Heavy Ions

The term *quasi-elastic heavy-ion scattering* refers to all processes in which the colliding nuclei do not undergo significant changes other than deviation from rectilinear motion. As a result of such processes, two other nuclei with masses close to the projectile and target masses and with kinetic energy close to the kinetic energy of the input channel are observed in the output channel. It is the latter condition that distinguishes these reactions from the processes of deep-inelastic scattering, considered in the next chapter.

Figures 4.4 and 4.5 show the energy spectra of the nuclei of krypton and oxygen in their collisions with lead nuclei at approximately the same energy of about 20 MeV/nucleon. For the first reaction, a magnetic spectrometer was used in combination with the time-of-flight system [63], which makes it possible to obtain

Fig. 4.4 The experimental spectrum of ^{86}Kr nuclei, measured in the ^{86}Kr + ^{208}Pb reaction at a beam energy of 18.2 MeV/nucleon and an angle of 12° in the laboratory system [63]. The inset shows the angular distribution of ^{86}Kr nuclei with energy loss in the range from 8 to 12 MeV. The dashed line is drawn for better visual effect

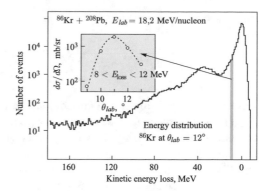

Fig. 4.5 The experimental spectrum of ^{16}O nuclei measured in the ^{208}Pb(^{17}O,^{16}O)^{209}Pb reaction at a beam energy of 376 MeV and an angle of 11.7° in the laboratory system [23]. The inset shows the angular distribution of ^{16}O nuclei following neutron transfer to the ground state of the ^{209}Pb nucleus

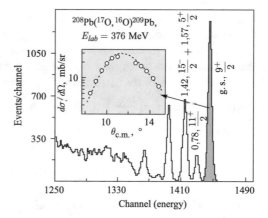

charge and mass resolution $\Delta Z/Z = 1/70$, $\Delta A/A = 1/260$ and reliably identify the final nuclei in the exit channel. However, a comparatively high energy resolution $\Delta E/E = 0.0026$ (that is, several MeV) still did not allow one to differentiate channels with fixed excited states of the outgoing nuclei in this reaction. However, this can be done in reactions with lighter nuclei, see Fig. 4.5. Despite this difference, in both reactions the angular distributions of the nuclei in quasi-elastic channels (that is, in channels with small energy loss and a small number of transferred nucleons) have a similar (bell-shaped) form with a maximum at some nonzero scattering angle (see insets in Figs. 4.4 and 4.5).

A careful analysis can help discover that this angle is close to the angle of the tangential collision, which is quite predictable. A physically clear explanation (and a quantitative description) of the bell-shaped form of the angular distributions of the quasi-elastic scattering processes of heavy ions can be obtained within the framework of a quasiclassical treatment of these processes. We write down the inelastic scattering cross section in the form $d\sigma_{fi}(\theta)/d\Omega = |f_{fi}(\theta)|^2$ (the quantity $f_{fi}(\theta)$ differs from $T_{fi}(\theta)$ only by a factor), and we expand the transition amplitude in terms of partial waves (as was done in Chap. 3 for elastic scattering), assuming that $\mu_f \approx \mu_i$, $k_f \approx k_i$, and that the transmitted angular momentum is much smaller than $l_{gr} = kb_{gr}$ (quasi-elasticity):

$$f_{fi}(\theta) = \frac{1}{2ik} \sum_{l=0}^{\infty} (2l+1) f_{fi}(l) e^{2i\chi_l} P_l(\cos\theta). \qquad (4.10)$$

Here $\chi_l = \delta_l + \sigma_l$ is the sum of the Coulomb and nuclear scattering phase shifts, and $f_{fi}(l)$ is the partial amplitude of the inelastic transition, that is, the probability of an inelastic process occurring along a trajectory with impact parameter $b = l/k$. If for the elastic channel ($f = i$), we put $f_{ii}(l) = 1$, then we obtain the elastic scattering amplitude at a nonzero angle, see expressions (3.19) and (3.27) (remember that for small l the phase shift δ_l has a large imaginary part due to the absorbing potential iW and $|e^{2i\chi_l}| \ll 1$).

However, for a quasi-elastic process, only a narrow range of impact parameters $b \sim b_{gr}$ contributes, since for $b < b_{gr}$ deep-inelastic reactions and fusion occur with great probability, and when moving along a trajectory with $b > b_{gr}$ the nuclei do not touch, and only elastic scattering occurs (see Fig. 3.11). The partial amplitudes of the transition can be approximated by the simplest expression $f_{fi}(l) \approx f_0 \exp\left[-\left(\frac{l-l_{gr}}{\Delta l}\right)^2\right]$. We expand the scattering phase $\chi_l = \delta_l + \sigma_l$ in powers $(l - l_{gr})$ taking into account Eq. (3.27):

$$2\chi_l \approx 2\chi(l_{gr}) + 2\frac{d\chi_l}{dl}(l - l_{gr}) = 2\chi(l_{gr}) + \theta_{gr}(l - l_{gr}).$$

For Legendre polynomials with large values of l, we use the approximate expression (3.26), valid for $l \gg 1, \theta > 1/l$:

$$P_l(\theta) \sim \frac{1}{2}\left(e^{i[(l+\frac{1}{2})\theta - \frac{\pi}{4}]} + e^{-i[(l+\frac{1}{2})\theta - \frac{\pi}{4}]}\right).$$

Finally, replacing the summation in Eq. (4.10) by integration with respect to l, we obtain the sum of two known integrals of the form

$$\int_{-\infty}^{\infty} x e^{i(\theta_{gr}\pm\theta)x} e^{-x^2/(\Delta l)^2} dx \sim e^{-(\Delta l)^2(\theta_{gr}\pm\theta)^2/4},$$

and the transition amplitude is reduced to the sum of two terms

$$f_{fi}(\theta) \approx -f_0\frac{\Delta l}{k} e^{2i\chi(l_{gr})}\left(\frac{l_{gr}}{2\sin\theta}\right)^{1/2} \times \left[e^{i[\lambda_{gr}\theta + \pi/4]} e^{-(\theta - \theta_{gr})^2(\Delta l)^2/4}\right.$$

$$\left. + e^{-i[\lambda_{gr}\theta + \pi/4]} e^{-(\theta + \theta_{gr})^2(\Delta l)^2/4}\right]. \qquad (4.11)$$

The first term in parentheses dominates and determines the bell-shaped form of the angular distribution in quasi-elastic reactions with heavy ions, the maximum being at $\theta = \theta_{gr}$. The width of such a distribution, $\Delta\theta = 2/\Delta l = 2/(k\Delta b)$, is inversely proportional to the localization of the process in the space of orbital angular momenta (or to the window of impact parameters, see Fig. 3.11), as it should be for any wave process.

Another visual description of the quasi-elastic scattering of heavy ions can be obtained, if we consider it as a certain probability of an inelastic transition in the motion of nuclei along classical trajectories that determine elastic scattering (*quasi-elasticity* just means a small perturbation when moving along an *elastic* trajectory). In this case, the scattering at an angle θ in an inelastic channel f is determined by the probability of elastic scattering at the same angle multiplied by the inelastic transition probability $i \to f$ along the trajectory leading to scattering at an angle θ, that is,

$$\frac{d\sigma_{fi}}{d\Omega}(\theta) \approx \frac{d\sigma_{el}}{d\Omega}(\theta) \cdot P_{fi}(\theta). \tag{4.12}$$

The cross section for the elastic scattering of heavy ions, $d\sigma_{el}(\theta)/d\Omega$, decreases rapidly when $\theta > \theta_{gr}$, as with small impact parameters that could lead to large angles of deflection, a strong overlap of the nuclei leads to a large loss of the kinetic energy of relative motion (see the total energy spectrum of the ^{86}Kr nuclei in Fig. 4.4, to processes of multi-nucleon transfer, fragmentation, or fusion (depending on the initial energy) and other reactions (see Sect. 3.6 and Fig. 3.11). At smaller angles (that is, large impact parameters), only elastic scattering occurs and the probability $P_{fi}(\theta \to 0) \to 0$. As a result, the cross section of the quasi-elastic process acquires a bell-shaped form. This qualitative explanation is well illustrated by Fig. 3.11, which shows the trajectories of incident oxygen nuclei in the reaction ^{16}O + ^{88}Sr and the angular distribution of ^{15}N nuclei formed as a result of the transfer of one proton to the ground state of the ^{89}Y nucleus. The probability of an inelastic transition, $P_{fi}(\theta)$, can be calculated, for example, in the quasiclassical approximation as an integral over time along the classical trajectory (leading to scattering at an angle θ) from the quantity $\langle f|\Delta V|i\rangle \exp\left[i/\hbar(E_f - E_i)t\right]$, where ΔV is the interaction responsible for the inelastic transition.

At collision energies close to the Coulomb barrier, the nuclei move practically along Coulomb trajectories (see Fig. 3.6), the impact parameter of the tangential collision becomes ever smaller and the angle $\theta_{gr} \to \pi$, that is, the maximum of the angular distributions for quasi-elastic processes shifts to backward angles, see Fig. 4.6a. In the conditions of the limiting periphery of the process, it can be assumed that the probability of an inelastic transition (for example, neutron transmission) should be maximum for the closest approach of the nuclei. For the Coulomb trajectory with an impact parameter b, the distance of closest approach $r_0(b) = \frac{\eta}{k} + \sqrt{\left(\frac{\eta}{k}\right)^2 + b^2}$. Since the scattering angle for a given impact parameter is determined by the relation (3.3), $\theta(b) = 2 \cdot \arctan(\eta/kb)$ or $b(\theta) = \frac{\eta}{k}\cot(\theta/2)$, then, for the given scattering angle, the point of closest approach of the nuclei is determined by the relation $r_0(\theta) = \frac{Z_1 Z_1 e^2}{E}\left(1 + \frac{1}{\sin\theta/2}\right)$. Thus, if the differential

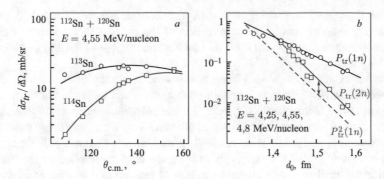

Fig. 4.6 (a) Experimental cross section for the transfer reaction of one and two neutrons in the collision of ^{112}Sn and ^{120}Sn nuclei with energy of 4.55 MeV/nucleon [69]. (b) The measured probability of transfer of one and two neutrons in the collision of ^{112}Sn and ^{120}Sn nuclei with energies of 4.25, 4.55, and 4.8 MeV/nucleon. The curves are drawn through the points. The arrow indicates the *gain factor* in the transfer of two neutrons

cross section of some inelastic process, $\frac{d\sigma_{fi}}{d\Omega}(\theta)$, is measured, it can be represented graphically by the function r_0. In order to unify such dependencies for reactions with nuclei of different masses, instead of the value r_0, the reduced distance of closest approach $d_0 = r_0/\left(A_1^{1/3} + A_2^{1/3}\right)$ is usually used. Using expression (4.12) from the cross section of a quasi-elastic process $\frac{d\sigma^{fi}}{d\Omega}(\theta)$, one can extract the transition probability $P_{fi}(d_0)$ as a function of the distance of closest approach.

As an example, Fig. 4.6a shows the cross sections for the transfer of one and two neutrons in near-barrier collisions of ^{112}Sn and ^{120}Sn nuclei, and Fig. 4.6b shows the corresponding probabilities of such transfers $P_{tr}(1n)$ and $P_{tr}(2n)$ as a function of the distance of closest approach of the nuclei. If they pass at a sufficiently large distances ($R > R_1 + R_2 + 2a$), then the probability of, for example, neutron transfer will be determined by the probability of finding the neutron outside the nucleus, that is, by the *tail* of its wavefunction. The neutron wavefunction outside the nucleus decreases exponentially and has the form $\varphi_n(r) \sim \exp(-\gamma r)/r$, where $\gamma = \sqrt{2\mu_n E_n^{sep}/\hbar^2}$ is determined by the binding energy (of the order of 8 MeV for medium and heavy nuclei). Thus, we should expect that the probability of neutron transfer in the peripheral reaction should also decrease exponentially with increasing distance of closest approach of the nuclei, that is, $P_{tr}(\theta) \sim e^{-2\gamma r_0(\theta)}$. This is indeed observed in the experiment, see Fig. 4.6b.

One would expect that the independent transfer of two neutrons in peripheral collisions is determined by the product of two probabilities, that is, $P_{tr}(2n) \approx P_{tr}^2(1n)$. Experimental data, however, indicate that $P_{tr}(2n) > P_{tr}^2(1n)$ (see Fig. 4.6). The *gain factor* of the probability of transfer of two neutrons indicates the possibility of their simultaneous transfer and turns out to be different for different combinations of nuclei. For the reaction considered here, $k_{2n} \approx 3$.

4.6 Reactions of Few-Nucleon Transfer

In some cases, an experiment is aimed not at transferring the nucleon to a particular state of the final nucleus, but at obtaining by nucleon transfer of a certain (sometimes exotic) isotope for its subsequent use or study (an example of such a reaction is shown in Fig. 4.6). Theoretical estimates of the cross sections for such processes are of much interest here. In the case of the transfer of a small number of nucleons, semiclassical analyses can be used to describe such processes [72, 73]. In this approach it is assumed that the colliding heavy ions move along classical trajectories (slightly distorted as a result of inelastic transitions), and the probabilities of nucleon transfers and collective excitations (vibrational states) are estimated within the framework of quantum models. The probability of nucleon transfer is determined using a parameterized microscopic form factor, taking into account the one-particle properties of the colliding ions and the average level-densities of single-particle states. Neutron and proton transfers and collective excitations are treated as independent events of low probability (for example, the transfer of two nucleons is certainly less likely than the transfer of one). This assumption is valid for tangential collisions. This approach was implemented in the computer code GRAZING [74], which was widely used in the analysis of the processes of few-nucleon transfer. Calculations using this program can also be done on-line at nrv.jinr.ru/\discretionary-nrv/\discretionary-webnrv/\discretionary-grazing.

In this model, for each impact parameter leading to scattering at a certain angle (see the previous chapter), there is a point of closest approach of the nuclei (turning point). It is assumed that inelastic excitation and nucleon-transfer processes occur mainly near this point, where the trajectory $r(t)$ can be approximated by a parabola and all the time integrals (for transition probabilities) can be calculated in explicit form. The differential cross sections are then calculated by summing overall impact parameters (orbital moments) corresponding to the scattering processes (it is assumed that the impact parameters, less than some critical value, lead to the fusion of the nuclei).

Figure 4.7 shows the experimental [17] and theoretical (calculated using the program GRAZING with standard values of the parameters) of the cross section of nucleon transfers in the reaction ^{40}Ca + ^{124}Sn at a beam energy of 170 MeV. The model of tangential collisions describes well the transfer of neutrons but greatly underestimates proton transfer. It can also be seen that in tangential collisions the cross section decreases exponentially with increasing number of transferred nucleons.

There is a very important question—which states of the final nucleus do the transferred nucleons occupy?. If the process of transfer of x nucleons occurs with the formation of final nuclei in their ground states: $a + A \rightarrow (a - x)_{g.s.} + (A + x)_{g.s.}$, then the energy $Q^x_{gg} = E_{bind}(a - x) + E_{bind}(A + x) - E_{bind}(a) - E_{bind}(A)$ is liberated (or lost), where E_{bind} is the binding energy of the corresponding nucleus. It is clear, however, that the transfer of nucleons can also occur to excited single-particle states lying above the Fermi energy. In this case, the kinetic energy of the nuclei in the exit channel is determined by the expression $E^f_{kin} \equiv E^i_{kin} + Q = E^i_{kin} + Q_{gg} - \varepsilon^*$, where ε^* is the excitation energy of the final nucleus. Figure 4.8

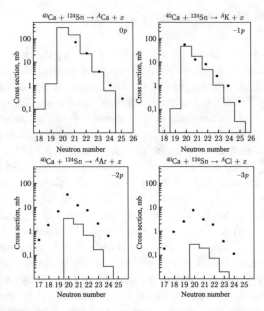

Fig. 4.7 Mass distributions of projectile-like fragments in channels with purely neutron transfers ($0p$) and in channels with transfer of 1, 2, and 3 protons in collisions of ^{40}Ca ions and ^{124}Sn nuclei at a beam energy of 170 MeV. The dots represent the experimental data [17], and the histograms show the cross sections obtained in the tangential collision model using the GRAZING program with the standard single-particle density parameters

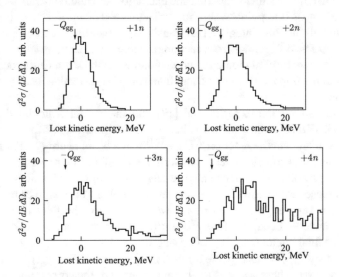

Fig. 4.8 Distribution of events by kinetic energy *loss* $E_i - E_f$ in channels with neutron transfers in the ^{58}Ni + ^{208}Pb reaction at a beam energy of 328 MeV [18]. The yield of nickel isotopes was measured at an angle of 90°

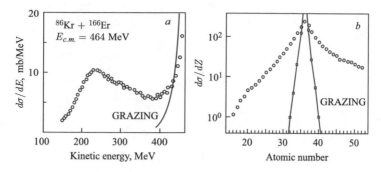

Fig. 4.9 Energy (**a**) and charge (**b**) distributions of reaction products in the collision of ^{86}Kr ions with ^{166}Er nuclei at 464 MeV in the center-of-mass system. The experimental data are taken from [27]

shows the experimental distribution of the kinetic energy *loss* (that is, the magnitude $E_{kin}^i - E_{kin}^f = -Q = -Q_{gg} + \varepsilon^*$) in the channels with transmission of 1, 2, 3, and 4 neutrons in the collision of ^{58}Ni ions with ^{208}Pb nuclei at a beam energy of 328 MeV [18] (the values below $-Q_{gg}$ are due to experimental errors). In this reaction, the transfer of neutrons to the ground state of the final nuclei proceeds with the release of energy: $Q_{gg}^{1n} = 1.6$ MeV, $Q_{gg}^{2n} = 6.3$ MeV, $Q_{gg}^{3n} = 6.0$ MeV, and $Q_{gg}^{4n} = 9.9$ MeV. However, as is evident from the experiment, in all channels the excited states are mainly populated with a Q-value close to zero; that is, the so-called *soft* transfer of neutrons from one mean field to another is realized, without an abrupt change in binding energy. As we shall see below, this conclusion is important for analyzing the role of the nucleon transfers in the reaction of sub-barrier fusion (see Chap. 5). We note that, for most combinations of colliding nuclei, few-nucleon transfer reactions can occur only with loss of kinetic energy, that is, $Q_{gg}^x < 0$ for such reactions, and the spectrum of kinetic energy loss initially begins with positive values (Fig. 4.9).

The model of grazing collisions certainly does not take into account processes with large kinetic energy loss that occur at smaller impact parameters (see, for example, Fig. 4.8a). As a consequence, this model is unsuitable for describing reactions involving the transfer of a large number of nucleons (Fig. 4.8b). The mechanisms of such reactions, called deep-inelastic scattering and quasi-fission, are discussed in the next chapter.

Chapter 5
Deep-Inelastic Scattering of Nuclei

As already noted, at low collision energies in reactions with heavy ions with nonzero impact parameters, a binary process dominates, in which two new fragments are formed: $a + A \rightarrow b + B$. In a narrow region of grazing collisions, it is possible to distinguish channels with a small transfer of mass, energy, and angular momentum, that is, the quasi-elastic scattering processes considered in the preceding chapter. However, in most cases, such collisions show a significant loss of kinetic energy, and sometimes a significant rearrangement of the charge and mass of the colliding nuclei. However, as we shall see below, even with a large loss of kinetic energy in the exit channel, nuclei close in mass and charge to the projectile and target are most likely to be observed. Such fragments are commonly referred to as projectile-like fragments (PLF) and target-like fragments (TLF).

If we use formula (4.6) for the amplitude of an inelastic process, then the cross section integrated over all angles, $\sigma_{fi}(E) = \int |f_{fi}(\theta)|^2 \sin\theta d\theta d\varphi$, simplifies due to the orthogonality of the Legendre polynomials:

$$\int P_l(\cos\theta) P_{l'}(\cos\theta) \sin\theta d\theta = 2/(2l+1)\delta_{ll'},$$

and we obtain

$$\sigma_{fi}(E) = \frac{\pi}{k^2} \sum_{l=0}^{\infty} (2l+1)T_{fi}(l, E) \equiv \sum_{l=0}^{\infty} \sigma_{fi}(l). \tag{5.1}$$

The values $T_{fi}(l)$ determine the probability of a transition to a final channel f for a given partial wave and, therefore, the partial cross sections $\sigma_{fi}(l) \leq \frac{\pi}{k^2}(2l+1)$. The quantity $\frac{\pi}{k^2}(2l+1)$ is the unitary limit of the partial cross section and is simply the cross section of the ring with the impact parameter b and width db: since $kb = l + 1/2$, then $\frac{\pi}{k^2}(2l+1)dl = 2\pi b db$ (see Fig. 3.2b).

© Springer Nature Switzerland AG 2019
V. Zagrebaev, *Heavy Ion Reactions at Low Energies*, Lecture Notes in Physics 963,
https://doi.org/10.1007/978-3-030-27217-3_5

Fig. 5.1 Schematic representation of the contribution of different partial waves to various reaction channels in the collision of heavy ions: fusion, deep-inelastic processes (DIP), quasi-elastic, and elastic scattering

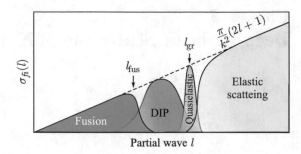

The mechanism of the nuclear reaction strongly depends on the impact parameter, that is, on l. In head-on collisions of light and medium-mass nuclei, the process of their complete fusion dominates (see below), whereas for large impact parameters, elastic scattering predominates. It is convenient that different reaction channels should be depicted by the corresponding partial cross sections as a function of l. Figure 5.1 shows a schematic representation of the partial cross sections for the processes of elastic, quasi-elastic, and deep-inelastic scattering of heavy ions and the process of their fusion, which dominates at small values of l. The relative contribution of different processes to the total cross section of the reaction

$$\sigma_{tot}^{R}(E) = \sum_{f \neq i} \sigma_{fi}(E)$$

strongly depends on both the nuclei involved in the reaction and the collision energy (in the collision of uranium nuclei, for example, the fusion cross section, that is, the cross section for the formation of a more or less spherical single compound nucleus is close to zero). At above-barrier energies for all combinations of heavy ions, deep-inelastic scattering processes make a significant (sometimes dominant) contribution to the total cross section of the reaction.

5.1   Experimental Systematics of Deep-Inelastic Scattering and Quasi-Fission

Figure 5.2 shows the angular distributions of the energy and charge of projectile-like fragments in the collision of ^{136}Xe at an energy of 1422 MeV (861 MeV in the center-of-mass system) with ^{209}Bi nuclei [75]. In the experiment, the energy and charge of projectile-like fragments were measured in the angular interval from $11°$ to $30°$ in the laboratory coordinate system with a step of $1°$. The three measured values (charge, angle, and energy of one of the fragments) were then used to restore the complete kinematics of the reaction in the center-of-mass system, assuming that only two nuclei with a Z/A ratio corresponding to the total system are formed in the exit channel. It was also assumed that the neutrons evaporated from the

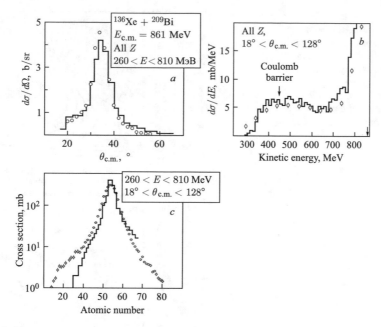

Fig. 5.2 The angular (**a**), energy (**b**), and charge (**c**) distributions of projectile-like fragments in the ^{136}Xe+^{209}Bi reaction at the beam energy of 1422 MeV [75]. The energy distribution was obtained by integrating over angles and summing over all charges. The angular distribution is shown for all fragments with a kinetic energy loss of no less than 50 MeV. The charge distribution was obtained by integrating over angles and energies. The histograms show theoretical calculations using the Langevin equations (see below)

excited nucleus (around one for every 12 MeV of the excitation energy) do not change the initial emission angle of this nucleus. In order to exclude the dominant elastic scattering channel, only events where the total kinetic energy of the resulting fragments was 50 MeV lower than the kinetic energy of the incident nucleus were taken into account.

In the process of deep-inelastic scattering, binary exit channels dominate, forming two fragments with masses close to the projectile and target masses. With increasing number of transferred nucleons, the cross section decreases monotonically (see Fig. 5.2c), but remains relatively large even when several tens of nucleons are transferred. The main feature of these processes is the presence of two components in the energy spectrum of the fragments formed, namely a quasi-elastic peak with energy close to the kinetic energy of the incident projectile and a wide energy distribution up to kinetic energies around the height of the Coulomb barrier of the colliding nuclei, see Fig. 5.2b. The *lost* kinetic energy (several hundred MeV in this reaction) clearly turns into the excitation energy of the fragments formed (which, despite this fact, retain their integrity). This phenomenon of dissipation of kinetic energy allows us, in principle, to speak of nuclear frictional forces (see below). Observation in the exit channel of nuclei with kinetic energy less than the

Fig. 5.3 The double
differential cross section for
the yield of potassium ions in
the ^{40}Ar + ^{232}Th reaction at
the beam energy of 388 MeV
(Wilczynski plot) [71]. On
the bottom, a schematic
explanation of the observed
energy-angle correlation is
given

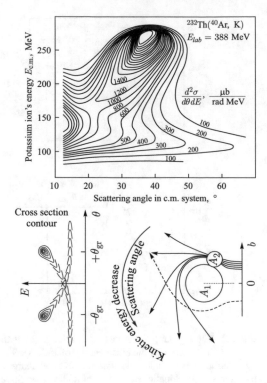

height of their Coulomb barrier can be explained only if we assume that their
dynamic deformation (elongation) at the moment of separation is large (similar
to the kinetic energy of strongly deformed fission fragments being equal to their
Coulomb energy at the scission point).

In many experiments on deep-inelastic scattering, a definite correlation is
observed between the emission angle and the value of the *lost* kinetic energy. The
clearest correlation of this type (called the Wilczynski plot [71]) is shown in Fig. 5.3,
where a double differential cross section is shown for the yield of potassium ions
in the ^{40}Ar + ^{232}Th reaction at the beam energy of 388 MeV. The cross section
has a pronounced maximum at the angle for grazing collisions and the energy of
the ions K, close to the energy of the initial beam (quasi-elastic process of few-
nucleon transfer). From this *peak* a ridge descends in the direction of decreasing
energy and the angle of departure, which at some point turns in the direction of
increasing scattering angle. This interesting behavior of the collision dynamics is
easily explained if it is assumed that the loss of kinetic energy is directly related
to the time during which the nuclei are in contact. Scattering at smaller angles
corresponds to longer contact times (see Fig. 5.3). The rotation of the inter-nuclear
axis by an angle greater than 180° leads to scattering to negative angles. Obviously,
to every trajectory like this corresponds a *symmetric* trajectory (shown by the
dotted line in Fig. 5.3) with a *negative* impact parameter, passing from the other
side of the nucleus and leading to scattering by a positive angle θ. In the energy-

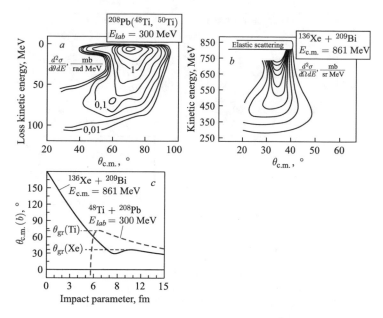

Fig. 5.4 (**a**) The double differential cross section for the yield of ^{50}Ti ions in the ^{48}Ti + ^{208}Pb reaction at the beam energy of 300 MeV [54]. (**b**) The double differential cross section for deep-inelastic scattering of ^{136}Xe nuclei on ^{209}Bi nuclei at the beam energy of 1422 MeV [75]. (**c**) The functions of the angle of deflection of the elastic scattering of ^{136}Xe + ^{209}Bi and ^{48}Ti + ^{208}Pb nuclei (the dotted curve), calculated with the proximity potential for nucleus–nucleus interaction

angle correlations, this manifests itself as a rotation of the maximum of the double differential cross section toward large angles.

The correlation between *kinetic energy loss* and *deflection angle* is less pronounced for some combinations of colliding nuclei, see Fig. 5.4a. In the collision of very heavy ions (for example, Xe + Bi) repulsive Coulomb forces dominate. The attractive nuclear forces prove insufficient to hold the nuclei in contact for a long enough time to deflect them to negative scattering angles, as happens for medium-mass nuclei. As a result, the angular distribution of products of deep-inelastic scattering, having a maximum for grazing collisions, becomes ever wider while the energy loss and the transferred mass increase (see Fig. 5.4b). In such heavy systems, complete fusion is impossible (or extremely small), and small impact parameters (see Fig. 5.4) also lead to deep-inelastic scattering or quasi-fission(see below). The total cross section of these processes is several barns and practically exhausts the total cross section of the reaction.

At low collision energies in the mass distribution of products of deep-inelastic scattering, instead of a monotonically decreasing cross section with increasing number of transferred nucleons shown in Fig. 5.2c, sometimes there is an increased yield of nuclei with closed shells in the region of intermediate masses $A_P < A < A_T$. Such processes are called *quasi-fission*, since, as is known, shell effects clearly

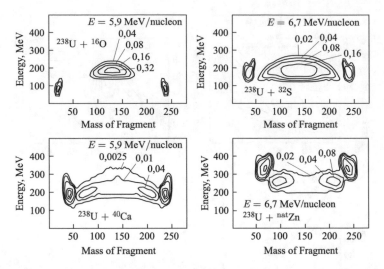

Fig. 5.5 Contour diagrams of the double differential cross section $d^2\sigma/dEdA$ (mb/MeV; values indicated near the curves) of the output of the products of the collision of ^{16}O, ^{32}S, ^{40}Ca, and ^{nat}Zn with ^{238}U nuclei at the low energies indicated in the figure [60]

manifest themselves in low-energy fission of heavy nuclei, leading, in particular, to an asymmetric mass distribution of fission fragments. Figure 5.5 shows the mass-energy distributions of reaction fragments formed during the collision of ^{16}O, ^{32}S, ^{40}Ca, and ^{nat}Zn with ^{238}U nuclei at low (near-barrier) energies.

In the collision of (low-mass) ^{16}O nuclei with ^{238}U nuclei, in addition to quasi-elastic scattering with the formation of projectile-like and target-like reaction fragments, the fusion of these nuclei occurs with high probability, forming an excited compound nucleus ^{254}Fm. The dominant channel for the decay of this nucleus is fission. This is why, in this reaction, fragments are observed with masses close to $A \sim (A_P + A_T)/2$; these come from the symmetric fission of excited fermium nuclei. Their total kinetic energy is much greater than the kinetic energy of the incident oxygen and corresponds to the standard value of the kinetic energy released in fission.

With increasing mass and charge of the incident nuclei, the mass distribution of the reaction products becomes wider. Along with the gradual decrease in the probability of formation of nuclei with masses close to $A_1 \sim A_2 \sim A_{CN}/2$ (indicating a decrease in the probability of fusion), an increased yield of nuclei in the mass range $A \sim 208$ is observed. This can obviously be attributed to the manifestation of shell effects (an increased probability of yields in the region of the doubly-magic nucleus ^{208}Pb). An explanation of this effect, observed for other combinations of colliding nuclei, is given below. Such behavior of the mass distribution of reaction products is not typical for deep-inelastic scattering (where a monotonic decrease in the cross section is observed, with increasing number of transferred nucleons, see Fig. 5.2c) and for the fission of an excited compound nucleus in which fragments of approximately equal mass are formed. Therefore, this process is called *quasi-fission*. Of course this is a question of terminology, and quasi-fission processes are simply a manifestation of shell effects in the low-energy processes of deep-inelastic scattering.

5.2 Potential Energy of Heavy Nuclear Systems, Diabatic and Adiabatic Driving Potentials

5.2.1 Nucleon Transfer and Driving Potentials

In slow collisions of nuclei, their behavior is governed mainly by the potential energy of their interaction. The observed regularities (see the previous section) allow us to say that an important role in such collisions is played by the charge and mass redistribution, as well as by the dynamic deformations of the colliding nuclei and their reaction products. Therefore, in addition to the dependence of the potential energy on the relative distance R between the nuclei, it is necessary to take into account the change in this energy with the dynamic deformations of the nuclei $\vec{\beta}_1$ and $\vec{\beta}_2$, (as one of the reasons, due to a change in the distance between the nuclear surfaces ξ for a fixed value R, see Sect. 2.2 and Fig. 2.8); and with the number of transferred nucleons from one nucleus to another changing. In the latter case, not only do the Coulomb and nuclear interaction energies change, but also their internal binding energies. It is convenient to include such a change in energy in the potential energy of the system. As a collective variable describing redistribution of nucleons between nuclei, a dimensionless quantity $\eta = (A_1 - A_2)/(A_1 + A_2)$, called the mass asymmetry, is usually chosen. For the diabatic interaction potential of separated nuclei we have the simple expression

$$
\begin{aligned}
V_{diab}(R, \vec{\beta}_1, \vec{\beta}_2, \eta) &= V_{12}(A_1, A_2; R, \vec{\beta}_1, \vec{\beta}_2) \\
&\quad + [M(A_1) + M(A_2) - M(A_P) - M(A_T)] \\
&= V_{12}(A_1, A_2; R, \vec{\beta}_1, \vec{\beta}_2) \\
&\quad + [B(A_P) + B(A_T) - B(A_1) - B(A_2)],
\end{aligned} \tag{5.2}
$$

in which $B(A)$ is the binding energy of the nucleus and the indices P and T refer to the projectile and the target (masses are expressed in energy units with the factor c^2 omitted).

The multi-dimensional potential energy written in the form (5.2) and describing not only the evolution of the nuclear system in the space of collective variables R and β, but also *pushing* it toward channels with positive reaction Q-values (that is, with the formation of more strongly-bound nuclei), is usually called a driving potential. The energy of the projectile and the target at infinity (in this case the

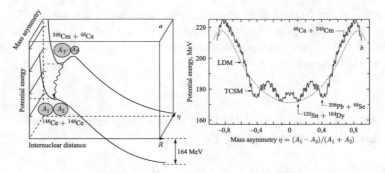

Fig. 5.6 (**a**) Potential energy in the entrance and exit channels in the reaction ^{48}Ca $+ ^{248}$Cm $\rightarrow ^{148}$Ce $+ ^{148}$Ce (in the entrance channel, mass asymmetry $\eta = 0.675$, and in the exit channel $\eta = 0$). (**b**) Potential energy in the nuclear system formed in the collision of ^{48}Ca and ^{248}Cm nuclei, depending on the mass asymmetry for a fixed distance between the centers of $R = 12$ fm nuclei (that is, approximately at the point of contact of the two nuclei) and with zero deformation. The solid and dashed curves are calculated in the two-center shell model (TCSM) and the liquid-drop model (LDM), respectively

expression in brackets turns to zero and $V_{12}(R \rightarrow \infty) \rightarrow 0)$ is chosen to be the zero level of energy.

As an example, Fig. 5.6a shows the potential energy in the entrance and exit channels in the reaction ^{48}Ca $+ ^{248}$Cm $\rightarrow ^{148}$Ce $+ ^{148}$Ce. For this reaction, $B(^{48}$Ca$)$ $+ B(^{248}$Cm$) - B(^{148}$Ce$) - B(^{148}$Ce$) = -164$ MeV, and thus the potential energy in the exit channel is lower than in the entrance channel, that is, after the contact of nuclei, nucleons are more likely to be transferred from the heavy nucleus ^{248}Cm to the lighter ^{48}Ca. Figure 5.6b shows the dependence of the potential energy of the two contiguous nuclei with zero deformation (and with the total number of nucleons A_1 $+ A_2 = 296$ and $Z_1 + Z_2 = 116$) as a function of redistribution of nucleons between them, that is, depending on the mass asymmetry η. This figure clearly shows the energy gain acquired by the system when the nuclei are *symmetrized* by mass, that is, the preferred evolution of the system towards $\eta \rightarrow 0$. Shell effects (the characteristic local minima in the potential energy) are discussed below.

5.2.2   Macro-Microscopic Model and the Adiabatic Potential Energy

In deep-inelastic scattering and quasi-fission processes, when a large number of nucleons are observed, the nuclei experience a strong overlap and the calculation of their potential energy is not so obvious. Figure 5.7 shows the evolution of the wavefunction of the *valence* neutron, which is initially in the state $2d_{5/2}$ in the ^{96}Zr nucleus as it approaches the ^{40}Ca nucleus at the near-barrier energy. The calculation is made in a realistic model with a time-dependent Schrodinger equation. It can be

^{96}Zr ^{40}Ca

Fig. 5.7 The change in the amplitude of the neutron wavefunction (the shaded area) initially located in the $2d_{5/2}$ state in the ^{96}Zr nucleus as it approaches the ^{40}Ca nucleus at a collision energy of 97 MeV in the center-of-mass system (approximately equal to the height of the Coulomb barrier)

seen that even before the nuclei touch (in fact, before they overcome the Coulomb barrier) the neutron wavefunction extends into the volumes of both nuclei. Thus, the valence neutrons (whose velocity inside the nuclei is much higher than the relative rate of approach of the colliding ions) begin to move in the field of the two nuclei even before they come into contact. As the nuclei approach and overlap, the same happens with other nucleons located in deeper single-particle states.

This analysis (and also many other ones carried out within the framework of microscopic models) shows that the low-energy collision of heavy ions has an adiabatic character (see Sect. 2.2), in which the independent motion of nucleons in the combined fields of initially separated nuclei is gradually rearranged to move in a general two-center mean field of the overlapping nuclei. The density of the nuclear matter and the volume of the nuclear system remain constant, and the shape of this system gradually changes from the configuration of two touching nuclei to that of a single more or less symmetrical nucleus (in the case of a fusion reaction), see Fig. 2.3.

To calculate the potential energy of such a system, in principle, a liquid-drop model of the atomic nucleus can be used. That is, in fact, the Weizsacker formula (B2) applied to a strongly deformed nuclear system in which the surface and Coulomb energy depend on the value and the nature of the deformation. The liquid-drop model, however, does not take into account shell properties that play an important role in the low-energy collisions of heavy ions. To account for these shell effects, the Strutinskii method [65] is usually used. Here the energy of a strongly deformed system is written as two terms: a *smooth* macroscopic part M_{mac} (calculated in one of the variants of the liquid-drop model) and a shell correction $\delta E = \sum_{i=1}^{A} \varepsilon_i - \left\langle \sum_{i=1}^{A} \varepsilon_i \right\rangle$. Here, $\varepsilon_i(R, \beta, \eta)$ are the energies of the single-particle states (shells) in the mean field of the nuclear system, and $\left\langle \sum_{i=1}^{A} \varepsilon_i \right\rangle = \int_{-\infty}^{\varepsilon_F} \varepsilon \tilde{g}(\varepsilon) d\varepsilon$—is the *smoothed* sum of the nucleon energies already taken into account in the macroscopic part M_{mac}. The adiabatic potential energy of a nuclear system, calculated in such a macro-microscopic approach, is written as follows

$$V_{adiab}(Z, A; R, \vec{\beta}, \eta) = M_{mac}(Z, A; R, \vec{\beta}, \eta)$$

$$+\delta E(Z, A; R, \vec{\beta}, \eta) - M(A_P) - M(A_T), \quad (5.3)$$

where the masses of the projectile and target are subtracted in order to set the potential energy of the nuclei at infinite distance in the entrance channel to zero. With a careful calculation, the diabatic (5.2) and adiabatic (5.3) potential energies coincide for the configuration of separated nuclei. This means that the adiabatic potential energy (5.3) can be used to describe both the fission process of the nucleus and the low-energy processes of deep-inelastic scattering and fusion.

As in the Weizsacker formula (B2) for the binding energy of the nucleus, the main terms in the macroscopic component of the mass of the nuclear system considered here

$$M_{mac}(Z, A; R, \vec{\beta}, \eta) \approx Zm_p + Nm_n - c_{vol}A - c_{surf}A^{2/3}B_{nuc}(R, \vec{\beta}, \eta)$$

$$-\frac{3}{5}\frac{Z^2e^2}{R_C}B_C(R, \vec{\beta}, \eta), \quad (5.4)$$

are the volume, surface, and Coulomb energies. Naturally, the latter two depend on the shape of the nuclear system determined by the collective variables $R, \vec{\beta}$, and η (in the case of a spherically symmetric mono-nucleus, the values B_{nuc} and B_C are equal to unity). The remaining terms that are necessary for a more accurate calculation of M_{mac}, as well as the optimal values of the model parameters, can be found in the specialized literature.

At relatively small deviations of the shape from a spherical mono-nucleus (for example, in fission processes), to calculate the shell correction, that is, the energies of the single-particle states $\varepsilon_i(R, \beta, \eta)$, the deformed mean field is usually used in the form of a Woods–Saxon potential. To describe the processes of deep-inelastic scattering and quasi-fission, in which the configuration of separated or slightly overlapping nuclei plays a large role, it is more efficient to use the so-called two-center shell model [43]. The essence of this model can be understood from Fig. 5.8. The mean-field potential in this model consists of two axially symmetric harmonic oscillators with independent centers. The two-center shell model correctly describes the transition from small ellipsoidal deformations $\delta_i = c_i/b_i - 1$ (see Fig. 5.8) near the ground state of the compound nucleus (where it coincides with the well-known Nilsson model) to strongly deformed forms up to the configuration of two separated nuclei. A more detailed description of the two-center shell model can be found on the NRV website [77], where the on-line calculation can also be made for the diabatic and adiabatic driving potentials for any nuclear system.

The multi-dimensional nature of the potential energy makes it very difficult to visualize, although one-dimensional graphs can give an idea of the dependence on a particular variable. The change in the potential energy of a system consisting of two

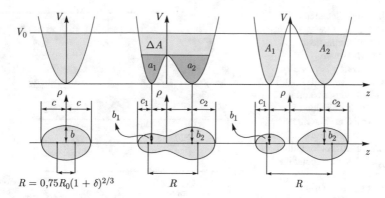

$$R = 0{,}75R_0(1 + \delta)^{2/3}$$

Fig. 5.8 Change in the shape of the nuclear system and the mean-field potential in the two-center shell model under ellipsoidal deformation $\delta_1 = \delta_2 = 0.5$ and initial mass asymmetry $\eta = 0.625$. The darker areas in the middle figure correspond to the still divided cores of the two nuclei, while ΔA denotes the number of already *collectivized* nucleons (obviously $a_1 + a_2 + \Delta A = A_1 + A_2$)

contiguous nuclei when the mass asymmetry is changed is shown in Fig. 5.6b. Deep minima due to the shell effects (that is, the formation of fragments in the region of doubly-magic nuclei ^{208}Pb or ^{132}Sn) are, in fact, valleys, taking into account the dependence of the potential energy on R as well (see Fig. 5.6a). In the case of three collective variables (the distance between centers, deformation, and mass asymmetry), the evolution of fission processes and deep-inelastic scattering occurs in the three-dimensional configuration space inside the cube, shown in Fig. 5.9. The change in the shape of the nuclei in the processes of fission, quasi-fission, and deep-inelastic scattering is shown schematically in the top part of this figure (the real trajectory, of course, lies inside this cube). Only the two-dimensional landscape of driving potential, that is, functions $V(R, \beta, \eta = 0)$, $V(R = R_{max}, \beta, \eta)$, and $V(R, \beta = 0, \eta)$ can be depicted on the sides of this cube (however, the reader must imagine the kind of potential energy inside the cube by himself).

Any physical system left to itself evolves primarily in the direction of diminishing potential energy. As will be shown below, the large *viscosity* of nuclear matter leads to a rapid loss of the kinetic energy of the relative motion of the colliding nuclei, which practically stop at the point of contact (the kinetic energy in this case becomes the excitation energy of the nuclei). The further evolution of the nuclear system is mainly determined by the landscape of its potential energy. The presence of deep minima (valleys) in the driving potential (see Figs. 5.6 and 5.9), caused by shell effects, leads to an increased yield of reaction fragments in these valleys. This is typical for the low-energy fission of nuclei and for the quasi-fission processes (see Fig. 5.5). The presence of internal excitation energy of the nuclear system (its temperature) leads to significant fluctuations in its evolution (see below) and the real trajectories, of course, are not as smooth as shown in the schematic diagram in Fig. 5.9.

Fig. 5.9 Top: Schematic representation of the evolution of the nuclear system in the space *elongation, mass asymmetry, deformation*. Bottom: Adiabatic driving potential relating to the collision of ^{48}Ca and ^{248}Cm. The figure shows the projections onto the plane of zero deformation of trajectories corresponding to the processes of deep-inelastic scattering (DIS), quasi-fission (QF), and fusion. The projection onto the plane of zero mass asymmetry ($\eta = 0$) is also shown for the trajectory of normal fission of the compound nucleus ^{296}Lv (Z=116) which passes through the saddle point and through the minimum of the isomeric state

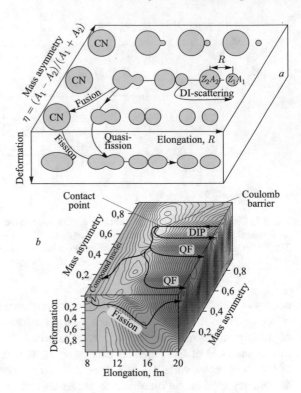

As the collision energy of the nuclei increases, their excitation energy increases (after contact). This leads to the erosion of the shell effects in the potential energy and in the mass yield of the reaction fragments. In the case of fission of a highly excited nucleus, symmetric distributions of the fragments dominate (see the global minimum at $\eta = 0$ in the potential energy of the liquid-drop model in Fig. 5.6b). On the other hand, in deep-inelastic scattering, projectile-like and target-like fragments dominate in the mass distributions with a monotonic decrease in the cross section as the number of transferred nucleons increases.

5.3   Transport Equations for Deep-Inelastic Nuclear Collisions: Frictional Forces

It is rather difficult to carefully describe the processes of deep-inelastic scattering of nuclei in which, in contrast to direct nuclear reactions, a large number of degrees of freedom play a role. Of the quantum approaches, the time-dependent Hartree–Fock method is sometimes used. This, however, still gives only a qualitative description of the dynamics of the collision (confirming, in particular, its adiabatic character) and is not used for quantitative description of experimental systematics. As was already noted above, the small de Broglie wavelength of the relative motion of heavy ions makes it possible to use the laws of classical mechanics to describe this motion. However, this is also difficult to do due to the large number of collective variables that play an important role in the collision process, because it is necessary to take account of the dissipation of the kinetic energy into internal excitation energy of the nuclei and because it is difficult to describe the processes of nucleon transfer in the framework of classical dynamics.

Experimental systematics indicate that in addition to the variable R describing the distance between the separated nuclei or the elongation of the nuclear system for the configuration of the mono-nucleus (see Fig. 5.8), an important role is played by nuclear deformations, $\vec{\beta}_1$ and $\vec{\beta}_2$, and for the description of nucleon transfers it is necessary to introduce a proton and neutron asymmetry, $\eta_Z = (Z_1 - Z_2)/(Z_1 + Z_2)$ and $\eta_N = (N_1 - N_2)/(N_1 + N_2)$. In describing collisions with nonzero impact parameter, which give the main contribution to deep-inelastic scattering, it is necessary to introduce the polar angle θ of the deviation of the trajectory from the beam axis, as well as the angles of rotation of the nuclei themselves, φ_1 and φ_2, shown in Fig. 5.10a.

Classical equations, namely the Langevin equations, take into account conservative and dissipative forces, as well as fluctuations of all collective variables caused by the internal motion of nucleons of highly excited nuclei. For any collective variable q (except for the charge and mass asymmetry) and its conjugate

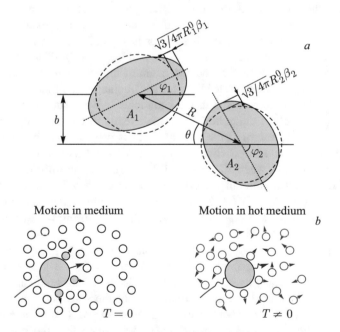

Fig. 5.10 (a) Variables used to describe the deep-inelastic scattering of heavy ions assuming a quadrupole dynamic nuclear deformation. (b) A schematic view of the *dissipative* motion of a body in a heated medium and the fluctuations of its motion

momentum, we can write a system of two equations

$$\frac{dq}{dt} = \frac{p_q}{\mu_q}, \frac{dp_q}{dt} = -\frac{\partial V}{\partial q} - \gamma_q \frac{p_q}{\mu_q} + \sqrt{\gamma_q T} \Gamma_q(t). \tag{5.5}$$

Here μ_q is the mass (inertial) parameter corresponding to the variable q (for the R variable this is simply the reduced mass in the case of separated nuclei, and for the rotation angle φ_i, for example, this is the moment of inertia of the i-th nucleus, etc.), γ_q is the friction coefficient responsible for the dissipation of the kinetic energy into the nuclear excitation energy $E^* = E_{c.m.} - V - E_{kin}$, $T = \sqrt{E^*/a}$ is the local temperature of the system, a is the level density parameter (see the statistical model of nuclear decay below), V is the potential energy of the system including the centrifugal term, and $\Gamma_q(t)$ is a normalized random variable with a Gaussian distribution, $\langle \Gamma(t) \rangle = 0$. Since the potential energy V is a function of the entire set of collective variables $\{q\} \equiv \{R, \beta, \eta_Z, \eta_N, \ldots\}$, then Eq. (5.5) is actually a system of a large number of coupled equations for these quantities.

The fluctuations that appear when a body moves in a heated medium are shown schematically in Fig. 5.10b. In this case the role of the medium is played by the internal motion of the nucleons. The kinetic energy of the collective motion (radial motion, surface vibration, and rotation of the nuclei) turns into the energy of their excitation. The difference in the character of the collective motion leads, in principle, to the difference of the corresponding frictional forces, and therefore, the coefficients γ_R, γ_β, and γ_θ are introduced. At sufficiently high excitation energies (several tens of MeV), this energy is more or less evenly (and rapidly) distributed over a large number of degrees of freedom, and the concept of temperature $T = \sqrt{E^*/a}$ can be introduced. The heated medium induces a reverse effect on the collective degrees of freedom, leading to their fluctuations, which are modeled by the inclusion of a random force in the equations. If we assume that the average value of the velocity of the collective variable tends to the thermal one, then the average (in absolute value) of the random force is proportional to the square root of the temperature and the friction coefficient. This means that during the initial stage of approaching (when $E^* = 0$ and $T = 0$), there are no fluctuations, and the trajectory of motion is only determined by the conservative forces. Excitation energy $E^* = E_{c.m.} - V(R, \beta, \eta) - E_{kin}$ and temperature are functions of collective coordinates and change together with them during the evolution of the system.

Inertial parameters $\mu_q(R, \beta, \eta)$ and frictional forces $\gamma_q(R, \beta, \eta)$ are also functions of the collective variables, and Eq. (5.5) are, in fact, a system of coupled equations. For overlapping nuclei, the dependence of these quantities on the coordinates can be calculated in the framework of the so-called Werner–Wheeler hydrodynamic model [19]. There is also a quantum approach to calculating inertial parameters, called the cranking model [22]. However, while the normal inertial parameter is clear

$$\mu_R(R \to \infty) = m_1 m_2/(m_1 + m_2),$$

and μ_β is determined in the liquid-drop model of the nucleus, the absolute values of the nuclear frictional forces (or nuclear viscosity for the configuration of the mono-nucleus) are not precisely defined. Apparently, the nuclear viscosity should also depend on the temperature of the system. Not all these questions have been completely studied and the phenomenological forces of friction are often used in the analysis of the experimental data.

The description of the process of nucleon transfer between colliding nuclei is the most difficult problem. On the one hand, the number of transferred nucleons is a purely discrete variable. On the other hand, we are talking about the transfer of a large number (dozens) of nucleons and for the description of such processes one can use *transport* approaches. The most suitable for this is the so-called master equation for the distribution function $\Phi(A, t)$

$$\frac{\partial \Phi}{\partial t} = \sum_{A'} \lambda(A' \to A)\Phi(A', t) - \lambda(A \to A')\Phi(A, t), \qquad (5.6)$$

where A is the number of nucleons in one of the nuclei (in the other one, there are $A_{CN} - A$) and $\Phi(A, t = 0) = \delta(A - A_P)$. The term $\lambda(A \to A')$ is the macroscopic probability of a transition from state A to state A' per unit of time. The macroscopic transition probability must be proportional to the number of possible states into which the transition occurs, that is, the density of levels. For this value, the symmetrized expression $\lambda(A \to A') = \lambda_0\sqrt{\rho(A')/\rho(A)}$ [45] has been suggested, where λ_0 is the nucleon transfer rate per unit of time, which, for a configuration with completely overlapping nuclear surfaces, should be of the order of the Fermi velocity divided by the linear size of the nuclear system, that is, $\lambda_0 \sim 10^{22}$ s^{-1}.

Unfortunately, Eq. (5.6) determines the evolution of the distribution function $\Phi(A, t)$ in time, rather than the value A itself or $\eta = (2A - A_{CN})/A_{CN}$. Thus, it cannot be solved in conjunction with the system of coupled Eq. (5.5), which at every step of integration requires the knowledge of a specific value of η (for calculating the driving potential, inertial parameters, and frictional forces), rather than the probability $\Phi(A, t)$ of finding it in a certain range of values. However, using certain rules, the master equation (5.6) can be transformed first into the Fokker–Planck equation

$$\frac{\partial \Phi}{\partial t} = -\frac{\partial}{\partial A}\left[D^{(1)}\Phi\right] + \frac{\partial^2}{\partial A^2}\left[D^{(2)}\Phi\right],$$

and then into the Langevin equation

$$\frac{dA}{dt} = D^{(1)} + \sqrt{D^{(2)}}\Gamma(t),$$

or, passing from the variable A to the variable $\eta = (2A - A_{CN})/A_{CN}$,

$$\frac{d\eta}{dt} = \frac{2}{A_{CN}}D^{(1)} + \frac{2}{A_{CN}}\sqrt{D^{(2)}}\Gamma_\eta(t). \tag{5.7}$$

This equation can now be added to the rest of the system of coupled Langevin equations (5.5), which determine the evolution of all collective variables in time. The equation for the mass asymmetry (derived from the corresponding master equation or from the Fokker–Planck equation) differs from the remaining equations in that it is written only for the variable η, and not for the conjugate momentum. This means that the transfer of nucleons is *inertia-free*.

The transport coefficients $D^{(1)}$ and $D^{(2)}$ are determined by the macroscopic transition probabilities $\lambda(A \to A')$:

$$D^{(1)} = \int (A' - A)\lambda(A \to A')dA',$$

$$D^{(2)} = \frac{1}{2}\int (A' - A)^2\lambda(A \to A')dA'.$$

If we assume that the nucleons are transferred from nucleus to nucleus mainly one by one (that is, $A' = A \pm 1$), then

$$D^{(1)} = \lambda(A \to A + 1) - \lambda(A \to A - 1),$$

$$D^{(2)} = \frac{1}{2}[\lambda(A \to A + 1) + \lambda(A \to A - 1)]. \tag{5.8}$$

Since, $\lambda(A \to A') = \lambda_0\sqrt{\rho(A')/\rho(A)}$ and $\rho(A) \sim \exp\left(2\sqrt{aE^*}\right)$, and $E^* = E_{c.m.} - V(R, \beta, A) - E_{kin}$, then, as is easy to see, $D^{(1)} \approx -\lambda_0\frac{1}{2T}\frac{\partial V}{\partial A}$ and $D^{(2)} \approx \lambda_0$, that is, the first term in Eq. (5.7) plays the role of a force pushing the mass asymmetry towards decreasing potential energy in the variable η or A (see Fig. 5.6b). The second term in Eq. (5.7) describes the diffusion of nucleons from one nucleus to the other, which dominates in the collision of symmetric masses or nuclei with closed shells, when $\frac{\partial V}{\partial A} \approx 0$ (see, for example, the charge distribution in the reaction ^{136}Xe $+ ^{209}$Bi in Fig. 5.2c).

Apparently, the transport coefficients $D^{(1)}$ and $D^{(2)}$ should also depend on the excitation energy of the nuclei, that is, on the temperature of the nuclear system. This issue has been studied rather insufficiently, and the experimental data are not enough to draw definite conclusions. For separated nuclei the probability of nucleon transfer decreases (see the previous chapter) and thus the nucleon transfer rate λ_0 should depend on R, tending to zero at large distances. To this end, in expressions (5.8) for macroscopic transition probabilities, one must use the expression

$$\lambda(A \to A \pm 1) = \lambda_0 \sqrt{\rho(A \pm 1)/\rho(A)} P_{tr}(R, \beta, A \to A \pm 1),$$

in which the probability of transfer of one nucleon for overlapping nuclei is $P_{tr} = 1$ and exponentially tends to zero when $R \to \infty$ (see Sect. 4.5 and Fig. 4.6b) from the previous Chapter).

For a more detailed description of the processes of deep-inelastic scattering and quasi-fission, and also for estimating the yields of different isotopes in such reactions (see below), one should separately consider neutron and proton transfers, and instead of one variable one must use neutron and proton asymmetry η_N and η_Z, that increases the number of variables and considerably complicates the calculations. The total number of variables that play an important role in the processes studied here is 8, namely $R, \theta, \varphi_1, \varphi_1, \beta_1, \beta_2, \eta_N$, and η_Z, the total number of equations is 14 (see Fig. 5.10a).

5.4   Calculation of Deep-Inelastic Cross Sections

Modern computers allow us to solve the system of coupled Eqs. (5.5) and (5.7) and, for a given impact parameter (for a given collision event), to find the *trajectory* of the evolution of the system in the multi-dimensional space of collective variables. In the exit channel such a trajectory leads to certain values of the charges and masses of the primary fragments formed in the reaction and scattered at certain angles with a certain kinetic energy and some excitation energy. These fragments are called primary fragments. In some cases (for small impact parameters) the colliding nuclei merge to form a compound nucleus with $Z_{CN} = Z_P + Z_T$ and $A_{CN} = A_P + A_T$ (see schematic in Fig. 5.11). The double differential cross sections for the yield of primary fragments Z'_1, A'_1 and $Z'_2 = Z_{CN} - Z'_1, A'_2 = A_{CN} - A'_1$, as in the experiment, are obtained by simply summing all the events that came into this channel

$$\frac{d^2\sigma(A'_1, Z'_1; E, \vartheta)}{d\Omega dE} = \int_0^\infty b db \frac{\Delta N(b; \eta'_Z, \eta'_N; E, \vartheta)}{N_{tot}(b)} \frac{1}{\sin \vartheta \Delta \vartheta \Delta E}. \tag{5.9}$$

Here ΔN is the number of events with formation of primary fragments of the reaction with the charge and neutron asymmetry η'_Z and η'_N having kinetic energy E in the interval ΔE and scattering angle ϑ in the interval $\Delta \vartheta$ for a given impact parameter b, and $N_{tot}(b)$ is the total number of events with this impact parameter. Because of the presence of fluctuations under the same initial conditions and a fixed impact parameter, different fragments with different energies are obtained in the exit channel. Therefore, for this impact parameter, it is necessary to *sample* as

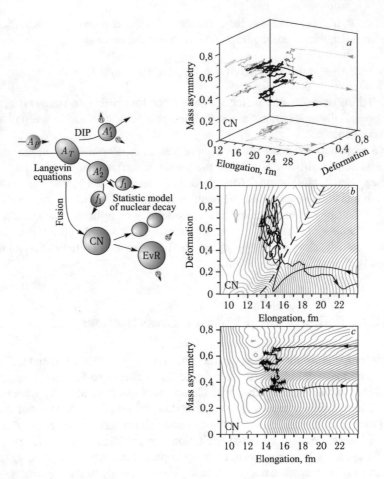

Fig. 5.11 On the left: A schematic representation of the collision of heavy ions leading either to the formation of excited primary fragments A'_1 and A'_2 or to fusion. In the second stage a statistical model of decay of the excited nuclei is used to obtain the observed products of the reaction. On the right: (**a**) One of the trajectories in the space *elongation, mass asymmetry, deformation* in the collision of an ion of ^{48}Ca with a ^{248}Cm nucleus at an energy $E_{c.m.} = 210$ MeV. (**b**) the projection of this trajectory onto the plane R, β (the dashed line denotes the position of the Coulomb barrier), and (**c**) projection onto R, η

many trajectories as possible in order to obtain an accurate cross section. As in the experiment, one needs to collect more statistics to see rare events. The undoubted advantage of this approach is the possibility of the simultaneous description of the processes of deep-inelastic scattering, quasi-fission, and fusion, that is, conservation of a kind of unitarity with all events resulting in one channel or another.

One of the trajectories in the *elongation, mass asymmetry, deformation* space is shown in Fig. 5.11 for the case of a ^{48}Ca ion colliding with a ^{248}Cm nucleus at the energy $E_{c.m.} = 210$ MeV. It is clearly seen how, after overcoming the Coulomb barrier and contact of the nuclei, a fast increase in their dynamic deformation occurs. This is due to the presence of a sufficiently deep minimum (potential pocket) in the potential energy of the system at $\beta = \beta_1 + \beta_2 \approx 0.6$ and $R \approx 16$ fm (see Fig. 5.11b). Subsequently a gradual transfer of nucleons from the heavy nucleus to the lighter one occurs with a decrease in the mass asymmetry of the system from its initial value $\eta \approx 0.675$ (^{48}Ca + ^{248}Cm) to $\eta \approx 0.4$ ($A_1 \sim 208$, $A_2 \sim 88$) at which point nuclear separation occurs. This tendency in the transfer of nucleons is also due to a decrease in the potential energy with decreasing η (see Figs. 5.11c and 5.6b).

As a rule the primary fragments of the reaction are formed in excited states and, therefore, such reactions are called dissipative. After the scattering, excitation energy is removed by evaporation of light particles and the emission of gamma quanta. Heavy primary fragments with high excitation energy can also undergo fission. For each event the probability of decay of the excited primary reaction fragments (that is, the emission of light particles competing with the process of sequential fission) is calculated within the statistical model (see below). As a result mass, energy, and angular distributions of the final reaction products are obtained to compare with those observed in experiments.

5.5   Analysis of Deep-Inelastic Scattering and Quasi-Fission

Figure 5.12a shows the experimental charge-energy distribution of reaction products in the collision of ^{86}Kr ions with ^{166}Er nuclei [18] at a center-of-mass kinetic energy of 464 MeV. One can clearly see the predominant yield of projectile-like ($Z \sim 36$) and target-like ($Z \sim 68$) fragments with a loss of kinetic energy of more than 250 MeV. In the range of kinetic energy of the emitted fragments of the order of 200 MeV, broad distributions of charge and mass are observed. After contact and release of this kinetic energy, the excitation energy of the system exceeds 200 MeV ($T > 2$ MeV). At this excitation energy all shell effects disappear (smooth out) and in the experiment one observes fairly smooth mass and charge distributions of the final reaction products (exponentially decreasing with increasing mass and charge transfer).

The distribution of fragment charge versus kinetic energy for the reaction ^{48}Ca + ^{248}Cm at a center-of-mass energy of 203 MeV [36] is shown in Figs. 5.13 and 5.14. Reactions of this type are of particular interest, since they make it possible to synthesize new superheavy elements through the complete fusion channel. As will be shown below that at low energies all more or less central collisions of light and medium-sized nuclei lead to complete fusion, that is, to the formation of a compound nucleus with an excitation energy determined by the collision energy and the binding energies of the participant nuclei, $E^* = E_{c.m.} + [E_{bind}(A_{CN}) - E_{bind}(A_P) - E_{bind}(A_T)]$. This compound nucleus can survive (that is, decay to the ground state of a long-lived evaporation residue) by emitting light particles and gamma rays or it can decay through the fission channel (for high excitation energy and/or low fission barrier height). By measuring the total

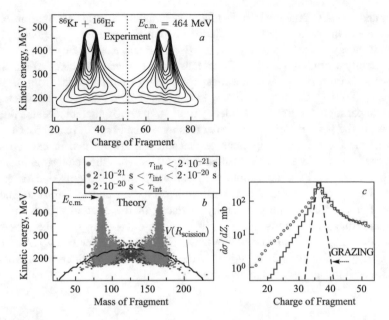

Fig. 5.12 The experimental [59] (**a**) and theoretical (**b**) charge-energy distribution of the reaction fragments from ^{86}Kr + ^{166}Er at $E_{c.m.}$ = 464 MeV. Theoretical events are divided into three groups (overlapping each other in the figure) depending on the reaction time: fast ($<2 \times 10^{-21}$ s), intermediate and long ($>2 \times 10^{-20}$ s). The curve on panel (**b**) corresponds to the potential energy of the system at the point of discontinuity (see text). Total charge distribution is shown on figure (**c**) together with theoretical calculations carried out within Langevin model (histogram) and performed by the GRAZING code (see Sect. 4.6)

yield of fission fragments and evaporation residues it is possible to obtain the total fusion cross section of nuclei at a given energy. For light and medium-mass nuclei at above-barrier collision energies, this cross section makes up a significant fraction of the total geometric cross section (several hundred millibarns at $E_{c.m.} > V_B$ and tens of millibarns at $E_{c.m.} \approx V_B$). In the collision of heavy nuclei their complete fusion (that is, formation of a more or less spherical mono-nucleus) occurs but with very low probability. Even for a head-on collision and a significant transfer of nucleons (indicating a long interaction time), the nuclei are most likely to separate again without forming a compound nucleus.

This process of quasi-fission, that competes with fusion, is easy to understand by looking at the potential energy of the system formed, for example, in the ^{48}Ca + ^{248}Cm reaction, see Fig. 5.14a. After overcoming the Coulomb barrier and making contact between the nuclear surfaces, the most energetically favorable evolution of the system (which has practically zero kinetic energy because of the large nuclear viscosity) is associated with the transfer of nucleons from the heavy nucleus (^{248}Cm) to the lighter one. Meanwhile the potential energy decreases since the nucleons in the light core are bound more strongly than in the heavy nucleus (see also Fig. 5.6). This leads to the appearance of a deep valley in the potential energy surface that

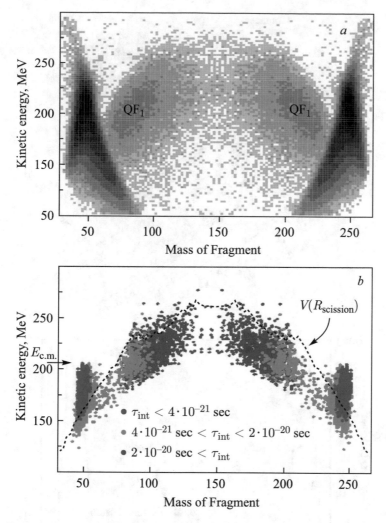

Fig. 5.13 Experimental [59] (**a**) and theoretical (**b**) charge-energy distribution for fragments from the reaction ^{48}Ca $+$ ^{248}Cm at $E_{c.m.} = 203$ MeV. The theoretical events are divided into three groups depending on the reaction time: fast ($< 2 \times 10^{-21}$ s), intermediate and long ($> 2 \times 10^{-20}$ s)

leads to formation of a doubly-magic ^{208}Pb nucleus in the exit channel and a lighter fragment complementing it. The movement along this valley, marked by the symbol QF$_1$ in Fig. 5.14a, is preferable to formation of a compound nucleus (the dotted line in the same figure).

The potential energy depends not only on the distance between the centers of the fragments and their mass asymmetry, but also on the deformation of the fragments, which strongly affects the evolution of the entire system (see Figs. 5.9 and 5.11). The calculations carried out using the Langevin equations, and taking into

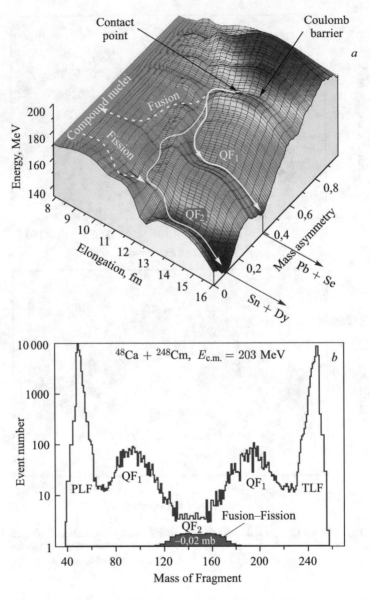

Fig. 5.14 The potential energy landscape is shown in panel (**a**) and the total mass distribution is shown in panel (**b**). PLF and TLF designate projectile-like and target-like fragments, and the symbols QF₁ and QF₂ designate asymmetric and mass-symmetric fragments from quasi-fission; see the corresponding schematic trajectories in panel (**a**)

account all the collective degrees of freedom listed above, are in good agreement with experimental systematics for quasi-fission (see Figs. 5.13 and 5.14). The *non-physical* experimental events in Fig. 5.13a, corresponding to the yield of projectile-like ($A \sim 50$) and target-like ($A \sim 250$) fragments with total kinetic energy exceeding the initial energy of 203 MeV, are caused by the technique of complex experimental measurements of the mass-energy distribution of the correlated quasi-fission products. This also explains the appearance of non-physical low-energy *tails* in the spectrum of fragments that are absent in the theoretical calculations.

In addition to the valley caused by the formation of lead-like fragments, a further valley is clearly visible in the potential energy surface in the mass-regions $A_1 \sim 132$ and $A_2 \sim 164$. This is caused by formation of the strongly-bound fragments in the region of the doubly-magic ^{132}Sn nucleus (see also Fig. 5.6). It is over this valley that the ordinary fission of the superheavy nucleus ^{296}Lv ($Z = 116$) (formed by complete fusion) occurs. These *fusion–fission* events are shown as a shaded zone in Fig. 5.14b. However, the fragments with such masses can also form in the process of quasi-fission without formation of the compound nucleus, see the schematic trajectory of QF$_2$ in Figs. 5.13 and 5.14. These fusion–fission and quasi-fission events are very difficult to separate experimentally. If this could be done, then the predictions of the cross sections for the formation of new superheavy elements (see below) would be greatly simplified. Theoretical estimates show that the total fusion cross sections for such heavy systems are hundreds of times smaller than those for light and medium-sized nuclei (of the order of 0.02 mb for the case of ^{48}Ca+^{248}Cm at the near-barrier energy $E_{c.m.}$=203 MeV).

5.6   Multi-Nucleon Transfer Reactions: Synthesis of Heavy Neutron-Rich Nuclei

In the reactions of deep-inelastic scattering and quasi-fission a large charge and mass transfer occurs from one nucleus to another. In recent years there has been particularly great interest in the process of multi-nucleon transfer, since other methods of producing and studying the properties of new nuclei and new elements have distinct limitations. Figure 5.15 shows the isotope map of all known chemical elements up to $Z = 118$. For light nuclei the most stable isotopes have an equal number of protons and neutrons. With increasing mass and charge the beta-stability line increasingly tilts toward the neutron axis, for example, the most stable isotope of uranium, ^{238}U, has 92 protons and 146 neutrons. Nuclei lying to the right of the beta-stability line have a neutron excess (β^- decay is possible for them) and to the left they have a proton excess and undergo β^+ decay.

Isotopes of light elements away from the stability line can be obtained by the process of fragmentation of heavier nuclei (see below). Neutron-rich isotopes of medium-mass nuclei are obtained in the fission of transuranic elements. Heavy nuclei can be obtained from the fusion reactions of lighter nuclei and in the chain of successive neutron captures followed by β^- decay, which increases the nuclear charge by one unit. The second process is realized in nuclear reactors, in nuclear explosions and naturally in supernova explosions. It was in this way that the

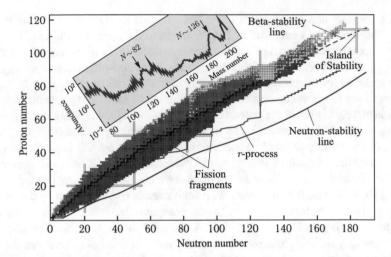

Fig. 5.15 Nuclear map. The inset shows the abundance of the elements found in nature. Note the characteristic sharp maxima in the region of strongly-bound nuclei, produced by the s-process, and wide maxima in the region of the *waiting points* ($N = 82$ and $N = 126$) of the r-process of astrophysical nucleosynthesis. The beta-stability line passes through the stable isotopes of nuclei (the dark squares) and is continued by a dashed line in the region of superheavy nuclei. The gray horizontal and vertical bars show the positions of the filled proton and neutron shells. The island of stability is thought to be located around the intersection of the closed shells $Z \sim 114$ and $N \sim 184$

elements heavier than uranium were artificially obtained. With the weak neutron fluxes realized in nuclear reactors, this process terminates on short-lived fermium isotopes, which experience spontaneous fission rather than β^- decay.

It is easy to see that when stable nuclei fuse, only proton-rich compound nuclei are formed. For example, upon fusion of the most neutron-rich isotopes of oxygen and tungsten, $^{18}O + {}^{186}W$, an excited neutron-deficient lead nucleus, ^{204}Pb, is formed which, after the emission of several neutrons, shifts even further from the beta-stability line towards the proton-rich isotopes. This is the reason why mainly proton-rich isotopes are synthesized for heavy (and superheavy) elements. All known isotopes of elements heavier than fermium ($Z = 100$) lie to the left of the stability line.

At the same time the properties of heavy neutron-rich nuclei are extremely important for understanding the natural origins of the elements heavier than iron, which cannot come from the fusion of lighter nuclei in ordinary stars. It is assumed that heavy elements arise as a result of neutron capture followed by β^- decay. Then, in order to explain the observed systematics in the production of stable isotopes of heavy elements and the existence of uranium and thorium in nature (separated from the continent of stable nuclei by elements with $Z = 84$–89 which do not have stable isotopes), it is necessary to assume that sources of very intense neutron fluxes exist in nature. Such fluxes can occur in supernova explosions and in the coalescence of neutron stars. In such high flows a stable isotope of a heavy element can capture several neutrons before β^- decaying, thus turning into a short-lived neutron-rich isotope, which will eventually β^- decay and produce elements of higher Z. This process of moving to the right and upwards on the nuclear map is called the r-process (rapid neutron capture).

The presence of nuclei with closed neutron shells (N = 50, 82, 126) should lead to the appearance of the so-called *waiting points* when the movement *to the right* stops. Indeed, the nucleus reaching the closed shell does not accept additional neutrons. Since the neutron-capture cross section becomes very low, it gives time for the nucleus to beta decay before a second capture. Once it has beta decayed, however, the neutron-capture cross section is large again and the system captures and jumps back to the magic-neutron number again but with Z increased by 1. This process repeats until we get closer to the stability line where the beta-decay lifetime becomes longer since the Q-value gets smaller as we move closer to the bottom of the mass parabola.

After the neutron flux disappears, all the neutron-rich nuclei located along the path through which the r-process passes undergo β^- decays and turn into stable nuclei (moving diagonally *left and upwards* on the map). It is the presence of the *waiting points* of the r-process that leads to the appearance of fairly wide peaks in the abundance of the stable elements marked in the inset Fig. 5.15. A complete understanding of the r-process environment (and its confirmation by appropriate numerical modeling) cannot be achieved unless we know the properties of the neutron-deficient isotopes of the elements involved.

The increased interest in the properties of heavy neutron-rich nuclei is also explained by the fact that the location of the single-particle states in such nuclei can change significantly due to the weakening of the nuclear spin–orbit interaction. This can also lead to the appearance of new magic numbers. This phenomenon has already essentially been proved for light neutron-rich nuclei. Unfortunately, heavy neutron-rich nuclei are not yet available for experimental study.

The process of multi-nucleon transfers in the low-energy collisions of heavy ions is perhaps the only method to obtain heavy neutron-rich nuclei. These processes are rather poorly studied, and the difficulty of the separation of multiple reaction products in charge and mass is only one of the reasons. The general pattern of such reactions shows an exponential fall in the cross section with an increase in the transferred charge and mass (see Figs. 4.7, 5.2, and 5.12). However, the stabilizing role of closed proton and neutron shells in the fragments generated in multi-nucleon transfer may significantly affect the probability of their formation.

In order to synthesize new neutron-rich nuclei lying below ^{208}Pb in the region of the filled neutron shell N = 126 and playing a key role in astrophysical nucleosynthesis (the last waiting point in the r-process), one could use proton transfer processes in the collision of ^{136}Xe with ^{208}Pb ions. Since such transfers occur while maintaining filled neutron shells in the projectile-like (N = 82) and target-like (N = 126) fragments, the reaction Q value remains positive or close to zero even when several protons are transferred from the heavy nucleus to the lighter

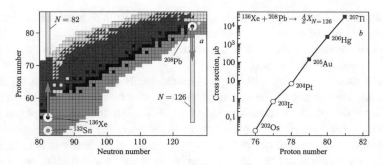

Fig. 5.16 Production of new neutron-rich nuclei with a closed $N = 126$ neutron shell in proton transfer in the collision of ^{136}Xe ions with ^{208}Pb nuclei at a center-of-mass energy of 450 MeV

one. In other words, such transfers are energetically favorable and the corresponding cross sections should not decrease too rapidly.

In Fig. 5.16 theoretical cross sections are shown for the formation of new neutron-rich nuclei located along the closed neutron shell $N = 126$ in the collision of ^{136}Xe ions with ^{208}Pb nuclei at the energy 450 MeV in the center-of-mass system. The main gain in this reaction stems from the fact that when protons are transferred from the target to the projectile, strongly-bound stable nuclei lying above ^{136}Xe are formed (see the left panel of Fig. 5.16). This leads to positive reaction Q values and to a sufficiently slow drop in the cross section with an increase in the number of transferred protons. An even greater gain in the cross sections for proton transfers can be expected in the reaction with a beam of radioactive tin nuclei, ^{132}Sn + ^{208}Pb, and also with the use of more neutron-rich transactinide projectiles or targets (for example, in the reactions of nucleon transfer from the target ^{238}U to incident ions of ^{196}Pt).

As we have seen above, shell effects play an even more prominent role in quasi-fission (Figs. 5.13 and 5.14) in low-energy collisions of heavy ions, leading in particular to an increased yield of nuclei with closed shells. This phenomenon can be used to obtain neutron-rich superheavy nuclei which cannot be synthesized in fusion reactions (see below). In the collision of heavy actinide nuclei (for example, U + Cm), one can expect an *inverse* (antisymmetrizing) quasi-fission process. Figure 5.17 shows the potential energy of a nuclear system formed upon the contact of ^{238}U and ^{248}Cm, depending on the number of nucleons transferred from one nucleus to the other. Obviously the transfer of nucleons from the uranium nucleus to the nucleus of curium, with the formation of ultimately strongly-bound nuclei in the lead region, is energetically favorable. The *lead* valley in this case is not as pronounced as for the system formed in the collision of ^{48}Ca and ^{248}Cm nuclei (see Fig. 5.6b) but here, as calculations show, a yield of nuclei in the lead region is also expected.

Since in low-energy heavy-ion collisions binary processes dominate, superheavy nuclei in the region $A \sim 280$ will naturally be complementary to the nuclei formed in the lead region. In the case of ^{48}Ca and ^{248}Cm, shell effects lead to an increased

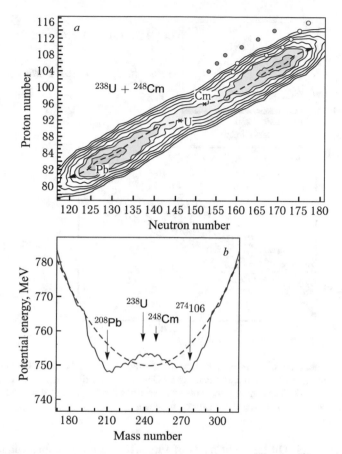

Fig. 5.17 The potential energy of the system formed on contact of ^{238}U and ^{248}Cm, depending on the number of nucleons transferred from one nucleus to the other: (**a**) as a function of Z and N and (**b**) as a function of the mass number A calculated along the dashed line in (**a**). The dashed line in (**b**) shows the potential energy calculated ignoring the shell correction δE in the formula (5.3). The circles in (**a**) indicate the positions of the superheavy nuclei synthesized in *cold* (filled circles) and *hot* (open circles) fusion reactions (see Chap. 6)

yield of nuclei with masses greater than the projectile mass and less than the target mass (see Figs. 5.13 and 5.14). Such a symmetrizing process was called (ordinary) quasi-fission. In the case of ^{238}U and ^{248}Cm, an increased yield is expected for nuclei that are lighter than the projectile (in the lead region) and heavier than the target (in the region of superheavy nuclei). Such an antisymmetrizing process can be called inverse quasi-fission.

Both of these processes are caused by shell effects. However, if in the first case the quasi-fission process plays a kind of destructive role, reducing the probability of fusion and formation of superheavy elements, in the second case, the process of inverse quasi-fission can, on the contrary, lead to an increased yield of neutron-rich

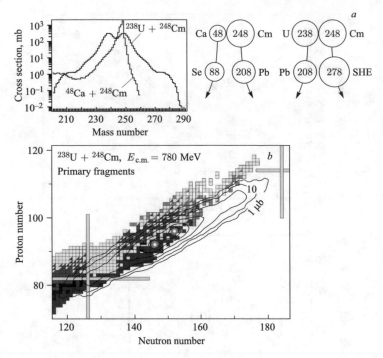

Fig. 5.18 (**a**) The process of *inverse* (antisymmetrizing) quasi-fission with the formation of neutron-rich superheavy nuclei in the collision of ^{238}U ions with a curium target at a center-of-mass energy of 780 MeV. (**b**) The contours show the cross sections for the yield of primary fragments in this reaction as a function of proton and neutron numbers

superheavy nuclei. On the right panel of Fig. 5.18 the topographical landscape is shown for the theoretical cross section for formation of primary fragments in multi-nucleon transfer reactions in the collision of ^{238}U and ^{248}Cm nuclei at a center-of-mass energy of 780 MeV. It is easy to see that with the help of such reactions we could come very close to the center of the island of stability, unattainable in the processes of nuclear fusion.

All primary fragments formed in the collision of heavy actinide nuclei have a significant excitation energy (depending on the collision energy), which, as a rule, exceeds the height of their fission barrier. The main channel for their decay is fission, and fission fragments will naturally dominate in the exit channel. However, at low collision energies a number (though small) of the superheavy nuclei can survive, emitting light particles (mostly neutrons) and gamma rays. In the dynamic calculations based on Eqs. (5.5) and (5.7), the excitation energies and angular momenta of all the primary fragments formed are known, and the probability of their survival can be calculated using the standard statistical model for the decay of atomic nuclei (see below). The results of such calculations are shown in Fig. 5.19. The cross sections for the yield of final nuclei fall by several orders of magnitude compared with the yield of the primary fragments. Nevertheless, these cross sections

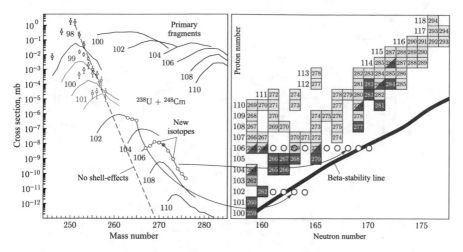

Fig. 5.19 Cross sections for the formation of superheavy evaporation residues (survivors) formed in the process of multi-nucleon transfer in the collision of ^{238}U ions with the target nuclei ^{248}Cm at a center-of-mass energy of 780 MeV. The experimental data for this reaction were taken from [59]

are of the order of 1000 pb for the synthesis of new neutron-rich isotopes of elements with $Z \leq 102$ and about 1 pb for the neutron-rich isotopes of Seaborgium ($Z = 106$) lying on the beta-stability line and, as was already noted, unattainable in fusion reactions. This cross section is in the experimentally achievable region.

Reactions of this kind have already been studied experimentally using radio-chemical methods of element separation [59]. The sensitivity of the experiment made it possible to reach the level of 1 μb, at which the isotopes of Mendelevium ($Z = 101$) were obtained, with a monotonic decrease with increasing charge for elements heavier than the target. A simple extrapolation of these data neglecting shell effects (which lead to an increased yield of nuclei in the lead region along with the complementary superheavy elements) is shown in Fig. 5.19 by a dotted line. This extrapolation leaves no chance of obtaining elements heavier than Rutherfordium ($Z = 104$) in multi-nucleon transfer reactions. Thus a further study of shell effects in low-energy collisions of heavy ions is extremely important.

Obtaining and studying the properties of new neutron-rich heavy nuclei is also complicated by the "technical" problem of their experimental separation from the numerous other products created by multi-nucleon transfers. It turned out that the existing methods and separators used to separate light products from fragmentation reactions or heavy evaporation residues from fusion reactions (see below) cannot be used to isolate and deliver to a heavy-isotope detector any element obtained by a multi-nucleon transfer reaction. In recent years, however, a combined method of separation based on stopping in a gas and subsequent, selective, resonant laser ionization has been proposed and intensively studied and developed. This method makes it possible to extract nuclei with a given charge, and the subsequent

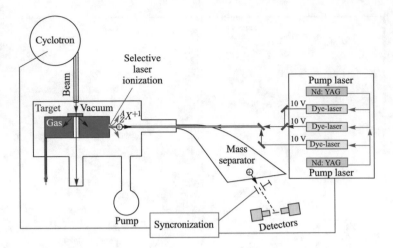

Fig. 5.20 Schematic view of the installation of resonant laser ionization of atoms of nuclear-reaction products stopped in gas with subsequent separation by mass and transport to the detection system

separation of singly-ionized isotopes by mass with a small magnet is possible. The lifetimes of heavy neutron-rich nuclei far exceed the transit time of such a facility.

A schematic diagram of an apparatus for extracting products of nuclear reactions by stopping them in a gas and their subsequent, selective, laser ionization is shown in Fig. 5.20. Neutron-rich isotopes of heavy elements are formed in the target by transfer reactions with heavy ions accelerated to energies of 5–10 MeV/A (depending on the projectile and target combination). The target is placed at the entrance to the gas cell (or inside its volume). The reaction products emitted from the target in the form of ions are decelerated and neutralized due to collisions with atoms of the buffer gas in a cell filled with high purity Ar or He. This gas, together with all the reaction products, is pumped to the vacuum through a supersonic nozzle or skimmer at the end of the cell. The wavelengths of the lasers are chosen to ionize (in a two or three-step process) the atoms of only a given nuclide at the nozzle exit. This is possible since atoms of different isotopes all have their own unique electronic state scheme and energies. The ions of the nuclide formed in this way are then accelerated to an energy of 30–60 keV and are separated magnetically in the mass separator. The resulting low-energy beam of ions has an extremely small emittance. The nuclei in the beam have a strictly defined mass and charge and are free of interference from other isotopes and isobars. This makes it possible, with high sensitivity, to study their decay properties (in a remote low-background laboratory). Recently, installations of this kind began to be built in several laboratories.

Chapter 6
Fusion of Atomic Nuclei

It has been found experimentally that with more or less central collisions at low (but above-barrier) energy, fusion, that is, the formation of a composite mono-nucleus with $Z_{CN} = Z_1 + Z_2$ and $A_{CN} = A_1 + A_2$, is likely to occur. Thus the fusion cross section at above-barrier energies is close to the geometric cross section. This is true, however, only for the collision of light and medium-mass nuclei (or low-mass nuclei with heavy ones). As the mass of the colliding nuclei increases, quasi-fission processes play an increasing role (see the previous chapter) and the probability of fusion sharply decreases, making the synthesis of new superheavy nuclei difficult (see below). At energies below the height of the Coulomb barrier, the fusion cross section decreases exponentially. The probability of fusion in this case is determined essentially by the probability of tunneling through a potential barrier (the inverse process to emission of an α particle or heavy cluster). The excitation of the collective degrees of freedom (for example, surface vibrations or the rotation of deformed nuclei) strongly influence the process of sub-barrier fusion (see below).

The compound nucleus formed by fusion has an excitation energy

$$E_{CN}^* = E_{c.m.} + E_{bind}(Z_{CN}, A_{CN}) - E_{bind}(Z_1, A_1) - E_{bind}(Z_2, A_2) \equiv$$
$$\equiv E_{c.m.} + Q_{fus}. \tag{6.1}$$

This excitation energy strongly depends not only on the center-of-mass kinetic energy of the collision, $E_{c.m.}$, but also on the binding energy of the colliding nuclei. When light nuclei fuse, the reaction Q-value, that is, $Q_{fus} = E_{bind}(Z_{CN}, A_{CN}) - E_{bind}(Z_1, A_1) - E_{bind}(Z_2, A_2)$, turns out to be greater than zero, since nuclei with medium mass have a large binding energy per nucleon (see Fig. 1.1). However, $Q_{fus} < 0$ when medium and heavy nuclei fuse. In this case fusion can occur only at collision energies $E_{c.m.} > |Q_{fus}|$, since the excitation energy of the compound nucleus cannot be negative. This restriction is generally not so important, since the threshold energy proves to be much lower than the height of the Coulomb

© Springer Nature Switzerland AG 2019
V. Zagrebaev, *Heavy Ion Reactions at Low Energies*, Lecture Notes in Physics 963,
https://doi.org/10.1007/978-3-030-27217-3_6

barrier, where the fusion cross section is in any case negligible because of the low penetrability of the barrier.

6.1 Detecting Fission Fragments and Evaporation Residues from the Compound Nucleus

It might seem that in order to measure the fusion cross section, it is necessary simply to measure the yield of composite nuclei (Z_{CN}, A_{CN}) formed in the reaction and recoiling forward (in the laboratory system). However, since these nuclei are in an excited state, they will decay on their way to the detector. The decay channels depend on the excitation energy and on the properties of the compound nucleus. At excitation energies of several tens of MeV for medium nuclei, the main decay channel is the emission (evaporation) of light particles (neutrons, protons, and α particles), and for heavy nuclei, it is the process of fission into two fragments of similar mass (see the next section). For nuclei with masses $A_{CN} \sim 200$ both processes (evaporation and fission) occur with almost identical probability. Thus, in order to measure the fusion cross section, $\sigma_{fus} = \sigma_{fis} + \sigma_{EvR}$, it is necessary to record the total yield of the evaporation residues and fission fragments, distinguishing them from the products of other reactions.

Figure 6.1 shows the dependence of the fusion cross section on the collision energy and the excitation energy of the compound nucleus in the reactions ^{48}Ca + ^{170}Er [57] ($Q_{fus} = -111$ MeV for this reaction and $E^*_{CN} = E_{c.m.} - 111$ MeV) and ^{16}O + ^{208}Pb [47] ($Q_{fus} = -46.5$ MeV). At high excitation energies, the compound nucleus ^{218}Ra formed in the first reaction mostly undergoes fission and, at excitation

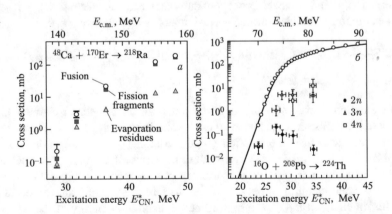

Fig. 6.1 Fusion cross sections (circles) in the reactions ^{48}Ca + ^{170}Er [57] (**a**) and ^{16}O + ^{208}Pb [47] (**b**) depending on the excitation energy of the compound nucleus, that is, on the collision energy. In figure (**b**) cross sections for the individual yields of evaporation residues and fission fragments are also shown

energies below 30 MeV, the survival probability of this nucleus (that is, removal of the excitation energy by emission of light particles and γ rays) is almost 50%. The heavier compound nucleus ^{224}Th formed in the second reaction has a lower fission barrier (equal to about 7.4 MeV, while in ^{218}Ra it is about 11 MeV) and the cross section for the formation of evaporation residues is much less than the total fusion cross section (see the right-hand side of Fig. 6.1).

To measure the yield of evaporation residues at forward angles, we must somehow separate them from the unscattered beam ions and from the products of other nuclear reactions. For this purpose various sorts of separators are used which first of all deflect the beam and then stop it in an appropriate detector. One can then select the ions with the desired mass A_{CN} (which can be determined from the time-of-flight and the energy released in the detector $A_{CN} \sim E \times \Delta t^2$) of ions moving with the speed of the compound nucleus $v_{CN} = (A_1/A_{CN})v_1$. A schematic view of an electrostatic separator that separates the recoiling (forward-moving) evaporation residues from the beam ions and from the products of other reactions due to their different deflection in a high-voltage electrictrostatic field is shown in Fig. 6.2 [10].

To accurately measure the fission cross section, one must determine the total yield of fission fragments flying at the angle 180° in the center-of-mass system and having masses of the order $A_{CN}/2$. In addition one needs to satisfy $A_{fis}^1 + A_{fis}^2 = A_1 + A_2$ within the accuracy of the mass resolution of the detectors, taking into account the possibility of evaporation of several neutrons, both from an excited compound nucleus (prescission neutrons) and from the fission fragments themselves (postscission neutrons).

It is obvious that for symmetric combinations of heavy nuclei (for example, Xe + Sn), the fusion cross section is virtually impossible to measure, since in this case the fission fragments cannot be readily distinguished from the products of elastic scattering and the processes of few-nucleon transfer.

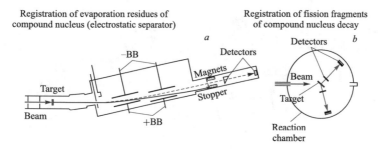

Fig. 6.2 Schematic view of an electrostatic separator [10] used to detect the evaporation residues formed in fusion reactions (**a**), and the reaction chamber (on the right) with two arms of the *start-stop* detectors for recording fission fragments from the compound nucleus (**b**)

6.2   Statistical Model for the Decay of an Excited Nucleus

As already noted, in fusion reactions a compound nucleus is formed, which, as a rule, has a rather high excitation energy. An exception is the radiative capture of light nuclei, in which only one or several γ rays are emitted. Such reactions are considered in the last section of this chapter.

The excitation energy of the compound nucleus E^* is determined by the expression (6.1), and its angular momentum J is determined by the impact parameter of the relative motion of the two nuclei for which the fusion occurred. Thus at a fixed beam energy, the excitation energy of the compound nucleus is fixed, and, in order to calculate the yield of evaporation residues and fission fragments, a summation should be made with respect to the angular momentum, taking into account the probability of fusion for each impact parameter.

The main channels of decay of the excited nucleus are the emission of light particles (primarily neutrons, protons, and α particles), γ rays and fission. These processes compete with each other and strongly depend on the charge and mass of the nucleus, its excitation energy and its angular momentum. The probability of each of the processes is determined using the so-called decay width. If we determine the total width of the decay of the (E^*, J) state of the nucleus Z, A as the inverse of its lifetime, $\Gamma(E^*, J) = \hbar/\tau$ ($\tau = T_{1/2}/\ln 2 = 1/\lambda$, where $T_{1/2}$ is the half-life and λ is the decay probability), then the *partial* decay widths for individual channels contribute as $\Gamma(E^*, J) = \Gamma_n + \Gamma_p + \Gamma_\alpha + \Gamma_\gamma + \Gamma_f + \cdots$

In some cases the partial widths can be measured experimentally. However, as a rule, these quantities are calculated within the statistical model using the concept of the nuclear level density $\rho_{ZA}(E^*, J)$ [34]. Partial decay widths for the emission by nucleus C (with excitation energy E^* and angular momentum J) of light particles $a(= n, p, \alpha, \ldots)$, or γ rays with multipolarity L, or for fission are determined by the following simple expressions

$$\Gamma_{C \to B+a}(E^*, J) = g^{-1} \int_0^{E^* - E_a^{\mathrm{sep}}} \sum_{l,j} T_{lj}(e_a) \sum_{I=|J-j|}^{I=J+j} \rho_B(E^* - E_a^{\mathrm{sep}} - e_a; I) de_a,$$

(6.2)

$$\Gamma_\gamma^L(E^*, J) = g^{-1} \int_0^{E^*} \sum_{I=|J-L|}^{I=J+L} f_L(e_\gamma) e_\gamma^{2L+1} \rho_C(E^* - e_\gamma; I) de_\gamma,$$ (6.3)

$$\Gamma_{\mathrm{fis}}(E^*, J) = g^{-1} \frac{\hbar \omega_B}{T} (\sqrt{1 + x^2} - x) \int_0^{E^*} T_{\mathrm{fis}}(e, J) \rho_C(E^* - e_a; J) de.$$ (6.4)

Here, $g = 2\pi \rho_C(E^*, J)$, $\rho_A(E^*, J)$ is the density of states of the nucleus A with the excitation energy E^* and spin J, $T_{lj}(e_a)$ is the probability to penetrate the Coulomb and centrifugal barriers for a light particle a with energy e_a.

When high-energy γ rays are emitted, dipole radiation with $L = 1$ usually dominates, and the strength function f_{E1} can be approximated by the expression [62]

$$f_{E1}\left(e_\gamma\right) = 3.31 \cdot 10^{-6} \left(\text{MeV}^{-1}\right) \frac{(A - Z)Z}{A} \frac{e_\gamma \Gamma_0}{\left(E_0^2 - e_\gamma^2\right)^2 + \left(e_\gamma \Gamma_0\right)^2}, \qquad (6.5)$$

with the resonance energy $E_0 = 167.23/(A^{1/3}\sqrt{1.959 + 14.074 A^{-1/3}})$ and width $\Gamma_0 \approx 5\,\text{MeV}$ for heavy nuclei.

For the fission width, Γ_{fis}, the Kramers correction is usually used. This takes account of the effect of nuclear viscosity η (see above) on the fission process, and the parameter $x = \eta/2\omega_0$, where ω_0 and ω_B are the characteristic frequencies (in the parabolic approximation for the nuclear potential energy as a function of deformation) for the ground-state configuration and for the configuration near the saddle point of the fission barrier (these values are close in magnitude, $\hbar\omega_0 \sim \hbar\omega_B \sim 1\,\text{MeV}$). Generally speaking, the nuclear viscosity should depend on the temperature $T = \sqrt{E^*/a}$ (where a is the level-density parameter, see below), and its value varies in the range of $(1 - 30) \times 10^{21}\,\text{s}^{-1}$. This quantity is rather poorly determined at present. The friction coefficient used above in the Langevin equations (5.5) is related to the viscosity by a simple relation $\gamma_q = \mu_q \eta$, where μ_q is the mass parameter of the collective variable q having dimensions of length, for example, the distance R between the centers of the fragments or the absolute change in the size of the nucleus upon deformation $s = R_0\beta$. Thus the friction coefficient has dimensions $[\text{MeV} \cdot \text{s} \cdot \text{fm}^{-2}]$.

With a parabolic approximation for the fission barrier, the penetrability $T_{\text{fis}}(e)$ in expression (6.4) for the fission width is determined by the formula

$$T_{\text{fis}}(e, J) = \frac{1}{1 + \exp\left\{-\frac{2\pi}{\hbar\omega_B}[e - B_{\text{fis}}(E^*, J)]\right\}},$$

where $B_{\text{fis}}(E^*, J) = B_0(E^*, J) - 1/2\hbar^2 J(J + 1)[1/\mathcal{J}_{g.s.} - 1/\mathcal{J}_{sd}]$ is the height of the fission barrier, taking into account the difference in the moments of inertia of the core in its ground state and for the configuration at the saddle point, $\mathcal{J}_i = k\frac{2}{5}MR^2(1 + \beta_{2,i}/3)$. The parameter $k \approx 0.4$ is an empirical correction factor that takes into account the decrease in the moment of inertia compared with a rigid body of mass M. The probability of tunneling under the fission barrier, that is, the penetrability $T_{\text{fis}}(e)$, is extremely small because of the large mass of the fission fragments. Therefore, it is often assumed that $T_{\text{fis}}(e < B_{\text{fis}}) = 0$ and the integration in (6.4) is performed from B_{fis} to E^*.

The height of the barrier itself, $B_0(E^*, J)$, is determined by its macroscopic properties for large deformations, that is, the change in the Coulomb and *surface* energies in the Weizsäcker formula when going from a spherical shape to a strongly deformed one at the saddle point, and the change in its microscopic properties (shell structure) under such deformations: $B_0(E^*, J) = B_{LD}(J) - \delta W(E^*, J)$.

In the liquid-drop model, the nuclear mass (that is, the sum of the nucleon masses minus the binding energy) can be simplified as follows (see Eqs. (1.1) and (1.2))

$$M(Z, A; \delta)c^2 = Zm_pc^2 + Nm_nc^2 - \left[c_{vol}A - c'_{surf}S(\delta) - Z^2 f(\delta) \right]. \quad (6.6)$$

When the nucleus deforms, its surface increases, so the number of less-bound surface nucleons increases and the total binding energy decreases (giving a larger mass). In formula (6.6) the parameter δ is intended to describe the change in the deformation: when $\delta = 0$, the nucleus has a spherical shape with a minimum surface area $S(0) = 4\pi R^2$. The Coulomb energy (proportional to the product of charges and inversely proportional to the distance between them) decreases with increasing deformation, since the average distance between protons increases. This is taken into account by the factor $f(\delta)$ in the last term in the formula (6.6), which decreases with increasing δ, $f(\delta = 0) = 3e^2/5R_C$ (see Eq. (1.2)).

Figure 6.3 schematically shows the change in the total energy of the nucleus, $M(Z, A; \delta)c^2$, as well as its surface and Coulomb energies with increasing deformation. It also shows schematically the change in the shape of the nucleus (Z, A) when it undergoes fission into two fragments (Z_1, A_1) and (Z_2, A_2). When these

Fig. 6.3 (a) A schematic representation of the change in the energy of the nucleus during its deformation and subsequent fission, calculated in the liquid-drop model [49]. (b) Fission barriers of isotopes of heavy nuclei located along the stability line. The dashed curve shows a calculation in the liquid-drop model [34], whereas the solid curve also accounts for shell corrections [44]. For some nuclei the experimental values of their fission barrier heights are shown [68]

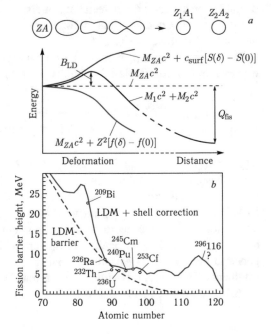

fragments fly off to large distance then, according to the law of energy conservation, $M(Z, A)c^2 = M(Z_1, A_1)c^2 + M(Z_2, A_2)c^2 + Q_{fis}$. The quantity Q_{fis} is the energy released in fission as kinetic energy of the fragments and their excitation energy (removed later by the evaporation of light particles and γ rays). For all heavy nuclei $Q_{fis} > 0$ and thus it is energetically favorable for them to split into two more strongly bound fragments. However, the nucleus needs to overcome the so-called fission barrier, B_{fis}. The height of this barrier is very large for nuclei with mass $A \sim 200$, and these nuclei are stable relative to fission (though they can fission if their excitation energy exceeds the barrier height). For heavier nuclei with $A \sim 240$, the height of the fission barrier is of the order of 6 MeV, and they can already experience spontaneous fission.

We note that the shape of the nucleus in fission is rather complex and cannot be described with just one parameter. The fission takes place in a multi-dimensional space of deformation parameters and is controlled by a multi-dimensional energy surface possessing several local minima and saddle points caused by the quantum (shell) properties of the system. None of these, however, changes the qualitative explanation of the fission mechanism shown in Fig. 6.3.

The liquid-drop model of the nucleus [49] predicts a monotonic decrease in the height of the fission barrier with an increase in the charge of the nucleus due to an increase in the Coulomb energy. However, the shell effects (arising from the quantum nature of the motion of nucleons in the nuclear mean field) leads to a significant change in this monotonic dependence and in particular, to a sharp increase in the height of the fission barrier for nuclei with closed shells. Figure 6.3 (bottom) shows theoretical estimates of fission barrier heights, made with and without taking into account the shell corrections for the ground-state energies of nuclei near the stability line. For some nuclei, the experimental values B_{fis} taken from [68] are also shown.

The microscopic correction $\delta W(Z, A; \delta)$ due to the irregular distribution of single-nucleon states in the nuclear mean field (see Fig. 2.2b) leads to many important effects.

(1) For nuclei with closed shells (or those which are close to them), the microscopic correction significantly reduces the ground-state energy, $\delta W(Z, A; \delta = 0) < 0$ and thus increases the height of the fission barrier (see the left part of Fig. 6.4).

(2) The dependence of the shell correction on deformation leads, in many cases, to the situation $\delta W(Z, A; \delta = 0) > 0$, so the mass of the nucleus is smallest for $\delta \neq 0$, that is, the nucleus has a non-zero deformation in its ground state.

(3) The negative values of the shell correction at $\delta \neq 0$ may lead to an additional minimum in the region of the fission barrier, which then becomes a *two–humped* barrier (see Fig. 6.4). Being at this minimum, a strongly deformed nucleus can have a rather long lifetime corresponding to isomeric states (in this case, the so-called shape-isomers).

(4) The shell correction leads to the stability of some very heavy nuclei (with respect to spontaneous fission) for which the macroscopic (liquid-drop) fission barrier practically disappears. These nuclei could not exist without the quantum properties of the nucleon motion (see the right-hand side of Fig. 6.3).

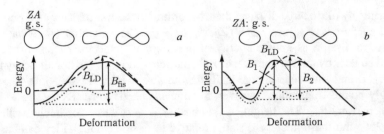

Fig. 6.4 A schematic representation of the dependence of the nuclear energy on deformation. The dashed curve shows the change in the macroscopic component of this energy; it is always minimum for zero deformation. The dotted curve is with a shell correction that can be negative at zero deformation (in this case the nucleus is spherical in its ground state) or positive (then the lowest value of the total energy occurs for non-zero deformation, as on the right-hand figure)

Experimental data indicate that the shell corrections must disappear at high excitation energy (high temperature). For this purpose, the so-called $\gamma_D \sim 0.06\,\text{MeV}^{-1}$ damping factor is usually used and the resulting fission barrier height is written as $B_0(E^*, J) = B_{\text{LD}}(J) - \delta W \cdot \exp(-\gamma_D E_{\text{int}})$, where δW is the shell correction for the ground state of the nucleus (and the shell correction at the saddle point is neglected). E_{int} is the *internal* excitation energy of the nucleus without the energy of rotation: $E_{\text{int}} = E^* - E_{rot}$, with $E_{rot} = (\hbar^2/2\mathcal{J}_{g.s.})J(J+1)$.

The density of excited nuclear states, which is the main part of Eqs. (6.2–6.4), has the form

$$\rho(E, J; \beta_2) = const \times K_{\text{coll}}(\beta_2)\frac{2J+1}{E^2}\exp\left[2\sqrt{aE_{\text{int}}(J)}\right], \qquad (6.7)$$

where $E = E^* - \delta$ ($\delta = 0$, Δ, or 2Δ for odd-odd, even-odd, and even-even nuclei, $\Delta = 11/\sqrt{A}$ MeV), and $K_{\text{coll}}(\beta_2)$ is an empirical factor increasing the nuclear level density due to the excitation of collective degrees of freedom (rotation of the nucleus or its surface oscillations) [37]. For the level-density parameter, a different parametrization is used that takes into account the damped shell correction

$$a = a_0\left[1 + \delta W\frac{1 - \exp(-\gamma_D E_{\text{int}})}{E_{\text{int}}}\right], \qquad (6.8)$$

where $a_0 = 0.073 \cdot A + 0.095 \cdot A^{2/3}\,\text{MeV}^{-1}$.

To estimate the cross section for evaporation-residue formation it is necessary to calculate the probability of the successive emission of the relevant light particles and γ rays that compete with fission (this process is called the evaporation cascade). The probability of emitting a light particle is determined by the simple relation $\Gamma_a/\Gamma_{\text{tot}}$, where

$$\Gamma_{\text{tot}}(Z, A; E^*, J) = \Gamma_n + \Gamma_p + \Gamma_\alpha + \cdots + \Gamma_\gamma + \Gamma_f$$

is the total width of the decay. When a light particle is emitted, the excitation energy of the remaining nucleus decreases by the kinetic energy of this particle and by its binding energy in the initial nucleus. For heavy nuclei (having a high Coulomb barrier that blocks the emission of charged particles), the main decay channels are neutron evaporation and fission. For light and medium nuclei, neutron evaporation competes with the emission of protons and α particles.

As an example, one can give an expression for the probability of the formation of a nucleus after the evaporation of x neutrons from a compound nucleus (Z, A) having initial excitation energy E_0^* and angular momentum J_0 (this case is important for reactions synthesizing superheavy elements, see below):

$$
P_{\mathrm{EvR}}(A \rightarrow A' + xn) = \int_0^{E_0^* - E_n^{\mathrm{sep}}(1)} \frac{\Gamma_n}{\Gamma_{\mathrm{tot}}}(E_0^*, J_0) P_n(E_0^*, e_1) de_1
$$

$$
\times \int_0^{E_1^* - E_n^{\mathrm{sep}}(2)} \frac{\Gamma_n}{\Gamma_{\mathrm{tot}}}(E_1^*, J_1) P_n(E_1^*, e_2) de_2 \cdots \times
$$

$$
\times \int_0^{E_{x-1}^* - E_n^{\mathrm{sep}}(x)} \frac{\Gamma_n}{\Gamma_{\mathrm{tot}}}(E_{x-1}^*, J_{x-1}) P_n(E_{x-1}^*, e_x)
$$

$$
G_{N\gamma}(E_x^*, J_x \rightarrow \mathrm{g.s.}) de_x. \tag{6.9}
$$

In this expression, $E_n^{\mathrm{sep}}(k)$ and e_k are the separation energy and kinetic energy of the k-th evaporated neutron, $E_k^* = E_0^* - \sum_{i=1}^k [E_n^{\mathrm{sep}}(i) + e_i]$ is the excitation energy of the nucleus after the evaporation of k neutrons,

$$
P_n(E^*, e) = C\sqrt{e}\exp[-e/T(E^*)]
$$

is the probability of evaporation of a neutron with kinetic energy e from the nucleus with excitation energy E^* (C is the normalization coefficient determined from the condition $\int_0^{E^* - E_n^{\mathrm{sep}}} P_n(E^*, e) de = 1$).

The quantity $G_{N\gamma}$ in (5.7) determines the probability that the excitation energy E_x^* and the angular momentum J_x remaining after emission of x neutrons will be carried away by γ rays. It can be approximated by the expression

$$
G_{N\gamma}(E_x^*, J_x \rightarrow \mathrm{g.s.}) = \prod_{i=1}^N \frac{\Gamma_\gamma(E_i^*, J_i)}{\Gamma_{\mathrm{tot}}(E_i^*, J_i)}, \tag{6.10}
$$

in which $E_i^* = E^* - (i - 1)\langle e_\gamma \rangle$, $J_i = J - (i - 1)$, where $\langle e_\gamma \rangle$ is the average energy of the dipole excitation (lying in the $0.1 \div 1$ MeV interval), and the number

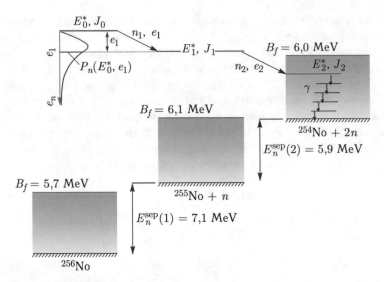

Fig. 6.5 Scheme of sequential decay of the excited nucleus ^{256}No with the emission of two neutrons and several γ rays. The notations are explained in the text

of emitted γ rays N is determined from the condition $E_N^* < B_{\text{fis}}$ (since at the excitation energies below the height of the fission barrier, the probability of fission is very small and $\Gamma_\gamma / \Gamma_{\text{tot}} \approx 1$).

The formula (6.9) and the quantities included in this formula are clearly illustrated by the decay scheme (Fig. 6.5) of the excited ^{256}No nucleus (formed, for example, in the ^{48}Ca + ^{208}Pb fusion reaction) at an excitation energy of the order of 20 MeV with emission of two neutrons and several γ rays and with the formation of a final evaporation residue ^{254}No. However, Fig. 6.6 shows as a function of excitation energy the decay widths and the survival probability of this nucleus when several neutrons are emitted. The height of the fission barrier is quite small ($B_{\text{LD}} = 1.26$ MeV, $\delta W = -4.5$ MeV, $E_n^{\text{sep}} = 7.1$ MeV), and so the fission channel dominates at excitation energies above 6 MeV. The probability of neutron evaporation is less by one or two orders of magnitude (depending on the excitation energy), and the probability of emission of charged particles (protons or α particles) is negligible. As the excitation energy decreases below E_n^{sep}, neutron evaporation is no longer possible and it is only γ-ray emission that competes with fission, completely dominating for $E^* < B_{\text{fis}}$.

Instead of a fairly time-consuming numerical calculation of several embedded integrals of the type (6.9), the so-called Monte Carlo method is often used to estimate the decay probabilities of a compound nucleus. This method simulates the evaporation cascade in competition with fission. For a given excitation energy, the first possible decay ($i = n, p, \alpha, \gamma, f$) is randomly chosen, allowing for its probability $\Gamma_i / \Gamma_{\text{tot}}$. If it is not a fission event, then the same is done for the next (daughter) nucleus with the new excitation energy reduced by the binding energy of

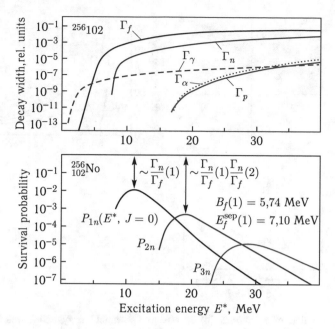

Fig. 6.6 The decay widths (in relative units) and the probability of survival of the excited nucleus ^{256}No in the channels with emission of 1, 2, and 3 neutrons

the emitted particle and its kinetic energy. Such an evaporation cascade is followed until the excitation energy becomes less than the height of the fission barrier. The final nucleus determines the specific decay channel for the original compound nucleus. This procedure is performed a large number of times and the probability of formation, for example, of a final, cold nucleus after the emission of x neutrons, y protons, and z α particles is defined as $N(xn, yp, z\alpha)/N_{tot}$, where $N(xn, yp, z\alpha)$ is the number of events of this kind, and N_{tot} is the total number of the events tested. In order to obtain at least one event with emission of a proton or an α particle in the decay of a ^{256}No nucleus with an excitation energy of 30 MeV, it is necessary to test at least $N_{tot} = 10^7$ events (see Fig. 6.6). Calculation of decay widths and the probabilities of formation of different final evaporation residues can be done directly on the Internet at http://nrv.jinr.ru/.

6.3 Fusion at Above-Barrier Energies

As already noted, when the surfaces of light and medium-sized nuclei come into contact with one another, it is very likely that complete fusion will take place, that is, the formation of a more or less spherical compound nucleus with $Z_{CN} = Z_1 + Z_2$ and $A_{CN} = A_1 + A_2$. So at above-barrier energies the fusion cross section is close to

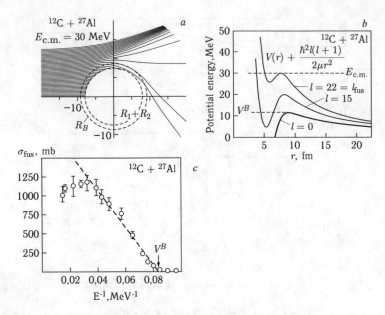

Fig. 6.7 (**a**) The field of trajectories, (**b**) the effective potential energy, and (**c**) the fusion cross section for the ^{12}C+ ^{27}Al reaction. In (**a**) the circles show the sum of the radii of the two nuclei (when this distance is reached, the nuclei will fuse) and the position of the Coulomb barrier

the geometric cross section. Figure 6.7a shows the trajectories of ^{12}C + ^{27}Al nuclei at a collision energy of 30 MeV, which is much higher than the Coulomb barrier $V_B = 11$ MeV. Figure 6.7b shows the potential energy of these nuclei including the centrifugal term $\hbar^2 l(l + 1)/2\mu r^2$. It is clear that without tunneling, the nuclei can come into contact for all partial waves less than some l_{fus} (in this case equal to 22), or equivalently, for all impact parameters less than $b_{\text{fus}} = l_{\text{fus}}/k$ $(\hbar^2 k^2/2\mu = E)$.

The cross sections for the yield of evaporation residues (following emission of x neutrons, y protons, and z α particles), for fission and for all fusion events can be written

$$\sigma_{\text{EvR}}^{xn,yp,z\alpha}(E) = \frac{\pi}{k^2} \sum_{l=0}^{\infty} (2l + 1) P(l, E) \cdot P_{\text{EvR}}^{xn,yp,z\alpha}(l, E), \qquad (6.11)$$

$$\sigma_{\text{fis}}(E) = \frac{\pi}{k^2} \sum_{l=0}^{\infty} (2l + 1) P(l, E) P_{\text{fis}}(l, E), \qquad (6.12)$$

$$\sigma_{\text{fus}}(E) = \sum_{x,y,z} \sigma_{\text{EvR}}^{xn,yp,z\alpha} + \sigma_{\text{fis}} = \frac{\pi}{k^2} \sum_{l=0}^{\infty} (2l + 1) P(l, E). \qquad (6.13)$$

For nuclei with a large mass, the probability of penetrating an effective barrier (hence the probability of contact of the surfaces) is $P(l \leq l_{\text{fus}}) \approx 1$, and $P(l >$

$l_{\mathrm{fus}}) \approx 0$. In that case

$$\sigma_{\mathrm{fus}}(E) \approx \pi/k^2 \sum_{l=0}^{l_{\mathrm{fus}}} (2l+1) \approx \pi l_{\mathrm{fus}}^2/k^2 = \pi b_{\mathrm{fus}}^2,$$

that is, the fusion cross section is very close to the *geometric* section, corrected for real, non-linear trajectories. The value l_{fus} can be easily estimated from the equality of the *effective* Coulomb energy barrier and collision energy (see Fig. 6.7b): $V^B + \frac{\hbar^2 l_{\mathrm{fus}}(l_{\mathrm{fus}}+1)}{2\mu R_B^2} = E$ or $l_{\mathrm{fus}}^2 = \frac{2\mu R_B^2}{\hbar^2}(E - V^B)$. Thus at above-barrier energies the fusion cross section is determined by a simple expression $\sigma_{\mathrm{fus}}(E) = \pi R_B^2 (1 - V^B/E)$. Representing the cross section as a function $1/E$ (see Fig. 6.7c), this expression can be used to determine the height of the Coulomb barrier for a given combination of nuclei. Linear extrapolation of this curve simply gives the height of the barrier. In reality, of course the fusion cross section does not become zero when $E < V^B$ (see the next section).

6.4 Sub-barrier Fusion: Hill–Wheeler Formula

The sub-barrier (and near-barrier) fusion of nuclei is of particular interest for two reasons. Firstly, in order to synthesize new elements (see below), it is necessary to choose as low a collision energy as possible and thus reduce the excitation energy of the formed compound nucleus. This increases the probability of its survival in competition with the dominant fission channel. Secondly, at such energies, the dynamics of the fusion process are much more complicated (see below), resulting in great interest in such studies.

Figure 6.8 shows the fusion cross sections for the reactions $^{16}O+^{154}Sm$ [42] and $^{36}S+^{90}Zr$ [64]. At sub-barrier collision energies, these cross sections do not become zero, but decrease exponentially, as they should if the possibility of quantum tunneling through the potential barrier is taken into account. If, however, we carefully calculate the probability of such tunneling for each partial wave $P(l, E)$ (this can be done simply by solving the corresponding radial Schrodinger equations with suitable boundary conditions), then on substituting the results obtained into expression (6.13), we find that the cross section obtained (shown in Fig. 6.8 with dotted curves) still differs significantly from the experimental data in the sub-barrier region.

In order to calculate fusion cross sections the simplest model was used at first, in which the colliding nuclei were assumed to be structureless and spherically symmetric. The nuclear part of the potential was chosen in the form of a Woods–Saxon potential with the parameters $V_0 = -105\,\mathrm{MeV}$, $r_0^V = 1.12\,\mathrm{fm}$, $a_V = 0.75\,\mathrm{fm}$ for the first reaction and $V_0 = -77.5\,\mathrm{MeV}$, $r_0^V = 1.15\,\mathrm{fm}$, $a_V = 0.8\,\mathrm{fm}$ for the second one. The heights of the corresponding Coulomb barriers, V^B, are indicated by the arrows in Fig. 6.8.

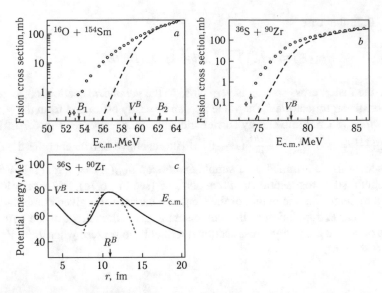

Fig. 6.8 Cross sections for fusion in the reactions (**a**) $^{16}O + {}^{154}Sm$ [42] and (**b**) $^{36}S + {}^{90}Zr$ [64]. Dashed curves show the results of calculations using the formula (6.13), taking into account quantum tunneling through the corresponding one-dimensional potential barriers. Figure (**c**) shows the interaction potential for $^{36}S + {}^{90}Zr$ and the approximation of the Coulomb barrier by a parabola (see text)

The considerable excess of the experimental fusion cross section in the sub-barrier region in comparison with the estimates obtained in this "one-dimensional" model (see Fig. 6.8c) is explained by the strong influence of the internal degrees of freedom of the nuclei (in this case, the orientations of the deformed ^{154}Sm nucleus and the surfaces vibrations of the nuclei ^{36}S and ^{90}Zr) on their relative motion (hence, on the fusion dynamics) at low energies. The mechanism of this influence is explained in the next section.

For a faster estimate of the penetrability of the barrier (without solving the corresponding Schrodinger equation), one can use the simple Hill–Wheeler formula [30], which is exact for the penetrability of a parabolic barrier. The real interaction potential of two nuclei (for example, shown in Fig. 6.8c) can be approximated in the barrier region by an inverted parabola:

$$V(r) \approx V^B + \frac{1}{2}V''(R_B)(r - R_B)^2,$$

where $V'' < 0$ at the maximum point. Such a parabolic potential is characterized by an oscillator frequency $\omega_B = \sqrt{-V''(R_B)/\mu}$ that depends on the width of the barrier (μ is the reduced mass of the colliding nuclei). In this case, the penetrability of the barrier is given by

$$P_{HW}(l, E) = \left[1 + \exp\left(\frac{2\pi}{\hbar\omega_B}[B(l) - E] \right) \right]^{-1}. \tag{6.14}$$

This formula takes into account an increase in the height of the effective barrier with increasing orbital angular momentum due to the centrifugal energy (see Fig. 6.6): $B(l) = V^B + \frac{\hbar^2 l(l+1)}{2\mu R_B^2}$. For deep sub-barrier energies one can easily take into account some broadening of the real barrier and a corresponding decrease in the value $\omega_B(E)$. In the sub-barrier energy region, this formula gives penetrability values very close to the exact values obtained by solving the Schrodinger equation.

6.5 Coupled Channels: Empirical and Quantum Description of Fusion

In the process of fusion of two atomic nuclei, their relative motion plays the most important role, since in order for them to form a compound nucleus, they need to overcome their Coulomb barrier and come into contact. However, as shown in Chap. 2 above, the interaction potential of two nuclei does not depend only on the separation, r, of their centers but also on the mutual orientation of these nuclei (if they have a static deformation in the ground state) and on the dynamic deformations of their surfaces (see Figs. 2.6, 2.7, and 2.8 and the corresponding formulas in Chap. 2).

This means that the one-dimensional model used above (dependence only on r) is too simplified. In fact as they approach, the nuclei must overcome at least a two-dimensional barrier (potential ridge, see Figs. 2.6 and 2.8). This barrier cannot be characterized by a single height, V^B, but is described by functions $V^B(\vartheta)$ or $V^B(\beta)$ depending on the orientation of the deformed nuclei or on their dynamic deformations. In Fig. 6.8a the arrows show the height of the Coulomb barrier for the spherical ^{16}O nucleus with the statically deformed ^{154}Sm ($\beta_2^{g.s.} \approx 0.3$). V^B indicates the height of the barrier assuming a spherical shape for ^{154}Sm, and B_1 and B_2 show the barriers for the limiting orientations of the ^{154}Sm nucleus: with the symmetry axis directed along the axis of separation of the nuclei (*nose-to-nose* configuration, $B_1 = V^B(\vartheta = 0)$), and perpendicular to it (*side-by-side* configuration, $B_2 = V^B(\vartheta = \pi/2)$).

The penetrability of a multi-dimensional barrier (that is, taking into account the coupling between the relative motion of the nuclei and their rotation and/or their dynamic deformations) can be estimated using the same Hill–Wheeler formula (6.14), providing that an appropriate *averaging* over the height of the barrier is performed. For the case of dynamic deformations (fusion of nuclei vibrating about a spherical shape, see Fig. 2.8), the resulting penetrability $P(l, E)$ that determines the fusion cross section (6.13) is calculated as

$$P(l, E) = \int F(B) P_{HW}[l, E; V^B(\beta)] dB, \qquad (6.15)$$

where $F(B)$ is a function normalized to unity, which can be approximated by a Gaussian

$$F(B) = N \cdot \exp\left(-\left[\frac{B - B_0}{\Delta_B}\right]^2\right) \qquad (6.16)$$

with a maximum at $B_0 = (B_1 + B_2)/2$ and width $\Delta_B = (B_2 - B_1)/2$. In this case, B_1 is chosen as the minimum height of the two-dimensional barrier V^{sd} (see the saddle point in Figs. 2.8 and 6.9), and B_2 corresponds to the Coulomb barrier of spherical nuclei, that is, $B_2 = V^B(\beta = 0)$.

In the case of statically deformed nuclei, the penetrability of the two-dimensional barrier must be simply averaged over orientations

$$P(l, E) = \frac{1}{4} \int\limits_0^\pi \int\limits_0^\pi P_{HW}[l, E; V_B(\vartheta_1, \vartheta_2)] \sin\vartheta_1 \sin\vartheta_2 d\vartheta_1 d\vartheta_2, \qquad (6.17)$$

where $V^B(\vartheta_1, \vartheta_2)$ is the orientation-dependent barrier height (see Figs. 2.6 and 2.7).

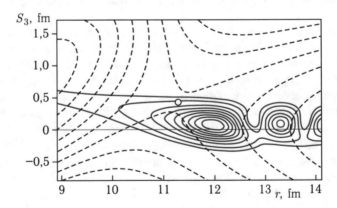

Fig. 6.9 The topographical landscape of the square of the modulus of the two-dimensional wavefunction (6.21) (solid lines) describing the process of fusion of ^{36}S and ^{90}Zr nuclei at $l = 0$ and energy $E_{c.m.} = 77$ MeV. The dashed curves show the landscape of potential energy. The circle denotes the position of the saddle point of the Coulomb barrier (the potential energy at this point V^{sd} is less than the height of the barrier at zero deformation). A similar surface for the potential energy of the interaction of ^{40}Ca and ^{90}Z nuclei is shown in Fig. 2.8

Formulas (6.15)–(6.17) together with expression (6.13) for the cross section realize a simple and intuitive empirical coupled channels (ECC) model of nuclear fusion, which perfectly describes the experimental data (see below). In the quantum channel-coupling (QCC) model for the sub-barrier fusion of atomic nuclei, it is necessary to solve a system of coupled Schrödinger equations corresponding to the total Hamiltonian describing the relative motion of deformed and rotating nuclei

$$
H = -\frac{\hbar^2 \nabla_{\mathbf{r}}^2}{2\mu} + V_C(r; s_{1\lambda}, \vartheta_1, s_{2\lambda}, \vartheta_2) + V_N(r; s_{1\lambda}, \vartheta_1, s_{2\lambda}, \vartheta_2)
$$

$$
+ \sum_{i=1,2} \frac{\hbar^2 \hat{I}_i^2}{2\mathcal{J}_i} + \sum_{i=1,2} \sum_{\lambda \geq 2} \left(-\frac{1}{2d_{i\lambda}} \frac{\partial^2}{\partial s_{i\lambda}^2} + \frac{1}{2} c_{i\lambda} s_{i\lambda}^2 \right).
\tag{6.18}
$$

In addition to the kinetic energy of the relative motion and the potential (Coulomb and nuclear) energy, this Hamiltonian also includes the kinetic energy of rotation of the nuclei ($\mathcal{J}_i$ are the moments of inertia of the nuclei, $i = 1, 2$) and the energy of their surface oscillations with multipolarity λ. The parameter $s_\lambda = \sqrt{(2\lambda + 1)/4\pi R_0} \cdot \beta_\lambda$ is the absolute elongation of the radius of the nucleus along the symmetry axis.

The intrinsic motion of the nuclear surfaces is described by the Hamiltonian,

$$
\hat{H}_{\text{int}}(\xi) = \sum_{i=1,2} \frac{\hbar^2 \hat{I}_i^2}{2\mathcal{J}_i} + \sum_{i=1,2} \sum_{\lambda \geq 2} \left(-\frac{1}{2d_{i\lambda}} \frac{\partial^2}{\partial s_{i\lambda}^2} + \frac{1}{2} c_{i\lambda} s_{i\lambda}^2 \right),
$$

where $\xi = \{\vartheta_i, s_{i\lambda}\}$ are the angles of rotation and deformation. The eigenfunctions of this Hamiltonian are well known (see Sect. 4.4): $\hat{H}_{\text{int}} \varphi_\nu(\xi) = \varepsilon_\nu \varphi_\nu(\xi)$. In the case of rotation $\varepsilon_\nu \equiv \varepsilon_I = \frac{\hbar^2}{2\mathcal{J}} I(I + 1)$ with eigenfunctions $\varphi_{IM}(\vartheta, \phi) \sim Y_{IM}(\vartheta, \phi)$, and in the case of harmonic oscillations of the surfaces $\varepsilon_\nu \equiv \varepsilon_n^\lambda = \hbar \omega_\lambda (n + 3/2)$ and $\varphi_n(s)$, are expressed in terms of the Hermite polynomials. Expanding the total wavefunction of the system over partial waves

$$
\Psi_{\bar{k}}(r, \theta, \xi) = \frac{1}{kr} \sum_{l=0}^{\infty} i^l e^{i\sigma_l} (2l + 1) \chi_l(r, \xi) P_l(\cos \theta),
\tag{6.19}
$$

we obtain the following equations for the partial wavefunctions

$$
\frac{\partial^2}{\partial r^2} \chi_l(r, \xi) - \frac{l(l + 1)}{r^2} \chi_l(r, \xi) + \frac{2\mu}{\hbar^2} \left[E - V(r, \xi) - \hat{H}_{\text{int}}(\xi) \right] \chi_l(r, \xi) = 0,
\tag{6.20}
$$

where $V(r, \xi) = V_C(r, \xi) + V_N(r, \xi)$. We now expand the functions $\chi_l(r, \xi)$ with respect to the complete set of wavefunctions describing the internal motion of the nuclei

$$\chi_l(r, \xi) = \sum_\nu y_{l,\nu}(r) \cdot \varphi_\nu(\xi). \tag{6.21}$$

The radial wavefunctions describing the relative motion in the channel ν satisfy a system of coupled, radial Schrodinger equations for the solution of which various mathematical schemes can be used,

$$y_{l,\nu}'' - \frac{l(l+1)}{r^2} y_{l,\nu} + \frac{2\mu}{\hbar^2} [E - \varepsilon_\nu - V_{\nu\nu}(r)] y_{l,\nu} - \sum_{\mu \neq \nu} \frac{2\mu}{\hbar^2} V_{\nu\mu}(r) y_{l,\mu} = 0. \tag{6.22}$$

Here, $V_{\nu\mu}(r) = \int \varphi_\nu^*(\xi) V(r, \xi) \varphi_\mu(\xi) d\xi$ are the matrix elements of the interaction (the off-diagonal elements are responsible for the transitions of the system from one channel to another, that is, for intrinsic excitations).

The boundary conditions necessary for solving this system of ordinary, coupled differential equations are formulated in order to describe the fusion process. At large distances we have an incident plane wave (describing the relative motion with momentum $\mathbf{k}$ in the input channel $\nu = 0$) and outgoing waves in all open channels $(E - \varepsilon_\nu > 0)$

$$y_{l,\nu}(r \to \infty) = \frac{i}{2} \left[h_l^{(-)}(\eta_\nu, k_\nu r) \cdot \delta_{\nu 0} - \left(\frac{k_0}{k_\nu} \right)^{1/2} S_{\nu 0}^l \cdot h_l^{(+)}(\eta_\nu, k_\nu r) \right]. \tag{6.23}$$

Here, $k_\nu^2 = \frac{2\mu}{\hbar^2} E_\nu$, $\eta_\nu = \frac{k_\nu Z_1 Z_2 e^2}{2E_\nu}$ is the Sommerfeld parameter, $\sigma_{l,\nu} = \arg \Gamma(l + 1 + i\eta_\nu)$ are the Coulomb phase shifts, and $h_l^{(\pm)}(\eta_\nu, k_\nu r)$ are the Coulomb partial wavefunctions with asymptotics $\exp(\pm i x_{l,\nu})$ where $x_{l,\nu} = k_\nu r - \eta_\nu \ln 2k_\nu r + \sigma_{l,\nu} - l\pi/2$ and finally $S_{\nu 0}^l$ are the elements of the scattering matrix.

As mentioned above, in the collision of light and medium-sized nuclei, their contact in the region behind the Coulomb barrier immediately leads to fusion (this is not so for heavy nuclei, see below). Thus we can still determine the fusion cross section by formula (6.13), in which the probability to penetrate the barrier in the channel ν is defined as the ratio of the flux that has reached a certain distance $R_{\text{fus}} \approx R_1 + R_2 < R_B$, to the incident flux $j_0 = \hbar k_0 / \mu$:

$$P(l, E) = -\frac{1}{j_0} \sum_\nu i \frac{\hbar}{2\mu} \left(y_{l,\nu} \frac{dy_{l,\nu}^*}{dr} - y_{l,\nu}^* \frac{dy_{l,\nu}}{dr} \right)_{r \leq R_{\text{fus}}}. \tag{6.24}$$

Currently we can find two widely used computational codes in which the above method of coupled channels for analyzing nuclear fusion is implemented: the Fortran program CCFULL [28] and the interactive program NRV Fusion, run directly in the browser window http://nrv.jinr.ru/. Figure 6.8 shows calculations of fusion cross sections performed within the empirical (ECC) and quantum (QCC) models of coupled channels in comparison with the experimental data obtained in the reactions $^{16}O + {}^{154}Sm$ [42] and $^{36}S + {}^{90}Zr$ [64].

The parameters of the interaction potentials of these nuclei are given in the preceding section. To describe the rotation of the deformed ^{154}Sm nucleus, the deformation parameters $\beta_2^{g.s.} = 0.3$ and $\beta_4^{g.s.} = 0.1$ were used, as well as the rotation energy $E_{2+} = 0.084\,MeV$. When describing the oscillations of a spherical ^{90}Zr nucleus, it was assumed that the largest contribution is made by the octupole vibrations of its surface ($\lambda = 3$, $\hbar\omega_\lambda = 2.75\,MeV$) with the amplitude of zero-point oscillations $\langle\beta_0\rangle = 0.22$. All these parameters are available from the database on the properties of atomic nuclei (http://nrv.jinr.ru/).

The form of the two-dimensional wavefunction $\chi_{l=0}(r, \xi)$ determined by formula (6.21) is shown in Fig. 6.9 for the fusion of ^{36}S and ^{90}Zr nuclei. In this case, $\xi = s_{\lambda=3}$, where $s_{\lambda=3} = \sqrt{(2\lambda + 1)/4\pi}\,R_0\beta_{\lambda=3}$ is the absolute value of the octupole deformation of the ^{90}Zr nucleus. At large distances the multi-channel wavefunction is concentrated in the region of small deformations $\beta_3 \approx 0$, which reflects the dominance of the zero-point oscillations of the nuclear ground state $\varphi_{\nu=0}(\beta)$ in the expansion (6.21). Oscillations of the modulus of the wavefunction at large distances are the result of interference between the incident and reflected waves.

It can be seen from Fig. 6.9 that at low energies (slow collisions) the nuclei experience noticeable deformation at the moment of contact. In this case, the Coulomb barrier is overcome mainly with positive values of deformation (elongation of the nuclei towards each other), leading to a decrease in the height of this barrier (see Fig. 2.8 and the potential energy landscape in Fig. 6.9 and, accordingly, to an increase in its penetrability. Quantum calculations provide a microscopic justification for the empirical channel-coupling models, in which simply averaging over the barrier height is used to calculate its penetrability, see formulas (6.15) and (6.16). Thus both the quantum and the empirical channel-coupling models, which take into account the effect of rotation and dynamic deformations on the fusion process (that is, on the process of overcoming the multi-dimensional Coulomb barrier) agree well with the available experimental data and explain well the increase in the probability of fusion in the sub-barrier energy region in comparison with the simplified model of a one-dimensional barrier penetrability.

6.6   Barrier Distribution Function

Experimental measurements of fusion cross sections are often so accurate that it becomes possible to calculate the energy derivatives of the function $\sigma_{\text{fus}}(E)$. The second derivative with respect to energy from the value $E\sigma_{\text{fus}}(E)$

$$D(E) = \frac{1}{\pi R_B^2} d^2 (E\sigma_{\text{fus}})/dE^2 \tag{6.25}$$

is directly related to the penetrability of the multi-dimensional barrier for the partial wave $l = 0$ [56]. Using formula (6.13) for the fusion cross section, we obtain the first derivative of the value $E\sigma_{\text{fus}}(E)$ (taking into account that $k^2 = 2\mu E/\hbar^2$):

$$\frac{d(E\sigma_{\text{fus}})}{dE} = \frac{\pi\hbar^2}{2\mu} \sum_{l=0}^{\infty} (2l+1) \frac{dP(l, E)}{dE}. \tag{6.26}$$

From formula (6.14) for the barrier penetrability it is seen that $P(l, E)$ depends on the combination of barrier height, angular momentum, and energy, that is, on the parameter $x = V^B + \frac{\hbar^2 l(l+1)}{2\mu R_B^2} - E$. Thus, $\frac{dP}{dE} = -\frac{dP}{dx} = -\frac{dP}{dl}\left(\frac{dx}{dl}\right)^{-1} = -\frac{dP}{dl} \frac{2\mu R_B^2}{\hbar^2} \frac{1}{2l+1}$ and

$$\frac{d(E\sigma_{\text{fus}})}{dE} = -\pi R_B^2 \sum_{l=0}^{\infty} \frac{dP(l, E)}{dl}.$$

In the collision of heavy nuclei, many partial waves contribute to the cross section of any reaction and $P(l, E)$ is a smooth function of l, thus the summation in (6.26) can be replaced by an integral over l. This is easily calculated, resulting in $d(E\sigma_{\text{fus}})/dE = \pi R_B^2 \cdot P(l = 0, E)$, or

$$D(E) = \frac{1}{\pi R_B^2} \frac{d^2(E\sigma_{\text{fus}})}{dE^2} = \frac{dP(l = 0, E)}{dE}. \tag{6.27}$$

In the classical case the barrier penetrability is a step function: $P(E > V^B) = 1$ and $P(E < V^B) = 0$, that is, $D(E) = \delta(E - V^B)$. In the quantum case the penetrability of the barrier is determined by a smooth function (6.14), varying from 0 to 1. Its derivative has a maximum at $E = V^B$ with a width $\Delta_B = \hbar\omega_B \ln(17 + 12\sqrt{2})/2\pi \approx 0.56\hbar\omega_B$ (for a parabolic barrier). Thus the second energy derivative extracted from experimental data (using finite differences) from the values $E\sigma_{\text{fus}}(E)$ (6.25) must have a narrow maximum at an energy equal to the height of the Coulomb barrier (which cannot be measured directly). It turned out that in many cases this function $D(E)$ has a richer structure in the form of additional maxima and minima (see Fig. 6.10). Its analysis allows us to extract

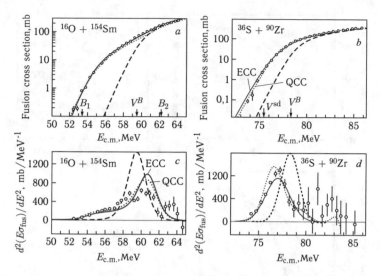

Fig. 6.10 Cross sections for fusion in the reactions $^{16}O + {}^{154}Sm$ [42] (**a**) and $^{36}S+{}^{90}Zr$ [64] (**b**). Dashed, dotted, and solid curves show the results of calculations performed in the one-dimensional barrier model (the same as in Fig. 6.8), and empirical (ECC) and quantum (QCC) models of coupled channels, respectively. The figures (**c**) and (**d**) show the corresponding barrier distribution functions (see Sect. 6.5)

detailed information on the excitations of the internal states of nuclei in the process of their fusion. The function $D(E)$ defined by Eq. (6.25) is usually called the barrier distribution function.

6.7 Neutron Transfer in the Process of Sub-barrier Fusion

In Sect. 5.2 above, it was shown that the interaction potential of two nuclei not only depends on the distance between their centers, their dynamic deformations and mutual orientation, but also on the transfer of nucleons between them (see Fig. 5.6). With this redistribution of nucleons, their binding energy (which is part of the total potential energy of the system) changes, leading to an increased yield of more strongly bound nuclei in the processes of multi-nucleon transfer and of quasi-fission (see Chap. 5). This redistribution occurs basically after the contact of the nuclear surfaces, that is, after they overcome the Coulomb barrier. It may seem then that it cannot influence the penetrability of the barrier and thus the fusion cross section.

However, it can be seen from Fig. 5.7 that, unlike protons, *external* (valence) neutrons being neutral can pass from nucleus to nucleus even before their surfaces come into contact and before they overcome the Coulomb barrier. Actually, it is with the collectivization of these valence neutrons, which begin to move in the

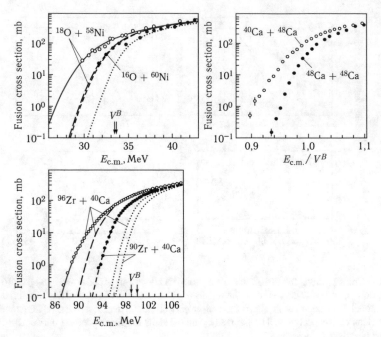

Fig. 6.11 Comparison of the fusion cross sections for close combinations of nuclei: $^{16}O+^{60}Ni$ and $^{18}O+^{58}Ni$, $^{48}Ca+^{48}Ca$ and $^{40}Ca+^{48}Ca$, $^{40}Ca+^{90}Zr$ and $^{40}Ca+^{96}Zr$. For each case in the first combination the transfer of neutrons is energetically unfavorable (all $Q_0(k) < 0$), and in the second combination the neutron redistribution occurs with energy release (see the text). The dotted curves show the cross sections obtained in the one-dimensional barrier model, and the dashed curves in the model of coupled channels with excitation of vibrational states of the colliding nuclei. The solid curves were obtained with allowance for the neutron transfer (see the text)

field (volume) of the two nuclei (see Fig. 5.7), that the process of fusion begins. It could be assumed (as was indeed done) that the probability of fusion of neutron-enriched nuclei should be enhanced. However, no such effect was observed. In itself, an excess of neutrons does not lead to an increase in the penetrability of the Coulomb barrier. For example, ^{40}Ca and ^{48}Ca nuclei fuse more readily in the sub-barrier region than ^{48}Ca and ^{48}Ca nuclei (see Fig. 6.11).

Neutron-rich isotopes have, as a rule, a slightly larger radius and this leads to a certain decrease in the height of the Coulomb barrier. However, this has very little effect on the cross section of sub-barrier fusion. A much greater effect was observed in those reactions where the transfer of neutrons from one nucleus to another occurs with energy gain, that is, with a positive value of Q-value (see Fig. 6.11).

Consider the process of nuclear fusion $A_1 + A_2 \rightarrow A_{CN}$, in which, at an intermediate stage, k neutrons are transferred from one nucleus to the other:

$$A_1 + A_2 \rightarrow (A_1 - kn) + (A_2 + kn) + Q \rightarrow A_{CN}. \tag{6.28}$$

We denote by $Q_0(k) = E_{bind}(A_1 - kn) + E_{bind}(A_2 + kn) - E_{bind}(A_1) - E_{bind}(A_2)$ the difference in the binding energies of the initial and intermediate nuclei in their ground states. Since in the transfer of neutrons, nuclei can be formed not only in the ground but also in an excited state, then $Q = Q_0(k) - \varepsilon^*$, where ε^* is the sum of the excitation energies of the nuclei $(A_1 - kn)$ and $(A_2 + kn)$, while $0 \leq \varepsilon^* \leq E$. For the vast majority of combinations of colliding nuclei $Q_0(k) < 0$ for $k = 1, 2, \ldots$, that is, redistribution of neutrons is energetically unfavorable.

However, in some cases $Q_0(k)$ turns out to have a positive value and in the intermediate stage in the *successive fusion* mechanism (6.28) we have a gain in energy, part of which can be used to excite $\varepsilon*$, and another part to get an additional kinetic energy of the relative motion of the nuclei $(A_1 - kn)$ and $(A_2 + kn)$. If the process of neutron redistribution occurs at distances before the Coulomb barrier, then this additional kinetic energy helps the nuclei to overcome the barrier. In this case, the successive fusion mechanism (6.28) with a positive value Q is a kind of *energy lift*, which helps the nuclei to overcome the Coulomb barrier. Redistribution of protons at large distances (at $r > R_B$) is much less likely due to their Coulomb barrier, leading to a faster decrease in the single-particle wavefunction of the proton outside the nucleus. Thus, the transfers of protons should have much less influence on the process of sub-barrier fusion of nuclei.

Figure 6.11 shows the cross section for the sub-barrier fusion for two close combinations of nuclei selected so that for one of them $Q_0(k) < 0$ for all k, and for the other $Q_0(k) > 0$. For example, in the combination $^{16}O + {}^{60}Ni$, any neutron transfers both from oxygen to nickel and backwards are energetically unfavorable, while the transfer of one neutron from ^{18}O to ^{58}Ni is accompanied by an energy release of $+0.95\,MeV$, and for the transfer of two $Q_0(2) = +8.2\,MeV$ neutrons. As a result, the fusion cross section of $^{18}O + {}^{58}Ni$ nuclei in the sub-barrier region is much larger than the $^{16}O + {}^{60}Ni$ fusion cross section (Fig. 6.11). Similarly for the combination $^{40}Ca + {}^{48}Ca$ $Q_0(2) = +2.6\,MeV$, $Q_0(3) = +0.14\,MeV$ and $Q_0(4) = +3.8\,MeV$, whereas for $^{48}Ca + {}^{48}Ca$, all $Q_0(k) < 0$. The same thing happens in the case of the fusion of ^{40}Ca with two different isotopes of zirconium. The fusion cross section of ^{40}Ca with ^{90}Zr (in this case all $Q_0(k) < 0$) is perfectly described in the model of coupled channels, taking account only of the vibrational properties of these spherical nuclei. On the other hand this model significantly underestimates the sub-barrier fusion cross section for $^{40}Ca + {}^{96}Zr$, for which neutron transfers from zirconium to calcium are possible with $+0.5, +5.5, +5.2, +9.6\,MeV$, etc. (See Fig. 6.11).

It is not easy to take into account the redistribution of nucleons within the scheme of quantum coupled channels, which further includes the excitation of rotational and vibrational states (see Sect. 6.5), and so far this has not been done. It is problematic because in the decomposition of the total wavefunction over excited states (6.20) and simultaneously over the states following a redistribution of nucleons (that is, over the states of different nuclei), a nonorthogonal and overfull set of basis functions must be used. Moreover, in medium and heavy nuclei the single-particle states (into which nucleons can be transferred) are distributed over a very large number of excited states (with corresponding spectroscopic factors, see Sect. 4.3), which are practically impossible to include in a usable microscopic scheme of coupled channels.

However, it is relatively easy to take into account the channels with redistribution of nucleons in the empirical model of coupled channels. In order to do this it is sufficient to use the relations (6.15) and (6.17) for the penetrability of the barrier in each of the channels with neutron transfer (6.28) by integrating over all possible Q values and summing over the number of transferred neutrons k. Let us denote as $\alpha_k(E, l, Q)$ the probability of k-neutron transfer at a center-of-mass energy E and orbital momentum l that corresponds to the formation of nuclei in the intermediate state with a certain excitation energy ε^* and the release (or loss) of energy $Q = Q_0(k) - \varepsilon^*$. For sequential neutron transfers, this probability can be estimated in the semiclassical approximation by means of the expression

$$\alpha_k(E, l, Q) = N_k \exp\left(-CQ^2\right) \exp\left(-2\kappa \left[r_0(E, l) - d_0\right]\right). \qquad (6.29)$$

Here, $\kappa = \kappa(\varepsilon_1) + \kappa(\varepsilon_2) + \cdots + \kappa(\varepsilon_k)$ where $\kappa(\varepsilon_i) = \sqrt{2\mu_n \varepsilon_i / \hbar^2}$, ε_i is the separation energy of the i-th transferred neutron, $r_0(E, l)$ is the distance of closest approach of the nuclei moving along a Coulomb trajectory with angular momentum l (see Sect. 3.1), $d_0 = R_1^{(n)} + R_2^{(n)} + 2\,\text{fm}$ and $R_2^{(n)}$ are the radii of the orbits of the valence (transferred) neutrons (the parameter d_0 is extracted from direct neutron-transfer experiments, see Sect. 4.5), and $N_k = \left\{ \int\limits_{-E}^{Q_0(k)} \exp(-CQ^2)dQ \right\}^{-1}$ is the normalization coefficient.

The total penetrability of the Coulomb barrier, taking into account the dynamic deformations of the nuclei or their rotation and possible redistribution of neutrons at the approach stage is still determined by the formulas (6.15) or (6.17), in which the probability of tunneling through the one-dimensional barrier is replaced by a value

$$\widetilde{P}_{\text{HW}}\left(l, E; V^B\right) = \frac{1}{N_{tr}} \sum_k \int\limits_{-E}^{Q_0(k)} \alpha_k(E, l, Q) \cdot P_{\text{HW}}(l, E + Q; V^B)\,dQ, \qquad (6.30)$$

where $N_{\text{tr}} = \sum_k \int \alpha_k(E, l, Q)dQ$ is the normalization coefficient and $\alpha_0 = \delta(Q)$.

The probabilities of neutron transfer (6.29) at sub-barrier energies are rather small (see, for example, Fig. 4.6b), and the first term with α_0, corresponding to the fusion process without neutron redistribution, dominates in the sum on k (6.30). Processes with neutron transfer make a noticeable contribution to (6.30) only if such transfers occur with positive Q values, that is, if $Q_0(k) > 0$ and the neutron transfer takes place in low-lying states. In this case a gain in penetrability is obtained due to the increased kinetic energy of the relative motion $E + Q$ and the penetrability increases exponentially $P_{\text{HW}}(l, E + Q)$ as this energy increases, see (6.14). A rapid decrease in the probability of neutron transfer with increasing k (see, Fig. 4.7) leads to the fact that only intermediate transfers of one or two neutrons with positive Q values have a significant effect on the process of sub-barrier fusion.

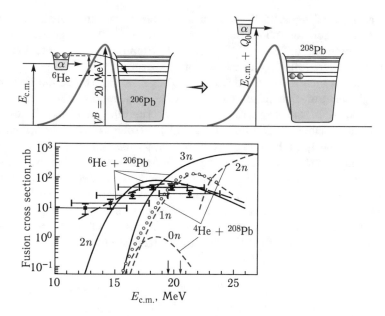

Fig. 6.12 Lower panel: cross sections for the yield of evaporation residues from the fusion of ^{6}He and ^{206}Pb nuclei (rectangles) [53] and ^{4}He + ^{208}Pb (circles) [6] with formation of the same compound nucleus ^{212}Po. For the ^{6}He + ^{206}Pb case, the dashed curve shows the cross section averaged over the ^{6}He beam energy (± 3 MeV in this complex experiment). Upper panel: the process of successive fusion of ^{6}He is schematically shown; loosely coupled neutrons pass to the nucleus ^{206}Pb (where their binding energy is greater than $Q_0(2) = 13.1$ MeV). This leads to an increase in the kinetic energy in the α particle + ^{208}Pb system (energy lift)

The redistribution of neutrons with positive Q values in the approach stage of fusing nuclei leads to a significant increase in penetrability of the Coulomb barrier, which agrees well with experiment. In Fig. 6.11 the solid curves show the fusion cross sections of the ^{18}O + ^{58}Ni and ^{40}Ca + ^{96}Zr reactions calculated using formula (6.30). The strongest effect from the intermediate neutron transfer with $Q > 0$ should be expected when weakly bound neutron-rich nuclei fuse with stable nuclei. Indeed the sub-barrier cross section for the fusion of ^{6}He and ^{206}Pb nuclei exceeds by several orders of magnitude the cross section for ^{4}He + ^{208}Pb (see Fig. 6.12).

In the case of ^{4}He and ^{208}Pb fusion, the intermediate neutron exchange is energetically unfavorable, while the transfer of two weakly bound neutrons from the ^{6}He nucleus to the ^{206}Pb can form the ground state of ^{208}Pb and release an amount of energy $Q_0(2) = 13.1$ MeV, comparable with a Coulomb barrier height of about 20 MeV. Of course, neutrons can also be transferred to excited ^{208}Pb states with a lower energy release (this is taken into account in formula (6.30) by integration over all possible values of Q). As a result the kinetic energy of the relative motion of the α particle and the ^{208}Pb nucleus increases and so does the corresponding probability of passing Coulomb barrier. This is exactly what is shown schematically on the upper part of Fig. 6.12. Theoretical predictions of the extremely large increase in the cross section for sub-barrier fusion of ^{6}He + ^{206}Pb (by several orders of magnitude compared to ^{4}He + ^{208}Pb), obtained from the successive fusion model (see Fig. 6.12) were fully confirmed by a later experiment [53].

A significant increase in the deep sub-barrier fusion cross section for weakly bound, light, neutron-rich nuclei may prove to be very important for astrophysical nucleosynthesis. In the standard scenario of primordial nucleosynthesis, it is assumed that unstable, weakly bound, light nuclei (such as ^{6}He, ^{8,9,11}Li, etc.) are formed with a fairly high probability in the (n, γ) reaction and then undergo β decay, turning into stable nuclei (see Fig. 6.13). At the same time, in the formation of heavier elements, a key role is played by reactions of deep sub-barrier fusion of light nuclei, the cross sections of which are extremely small because of the low kinetic energy of the relative motion inherent in astrophysical processes. In this connection, the inclusion of the reactions of fusion of light, loosely bound nuclei into the general scheme (though not very many are formed) can significantly

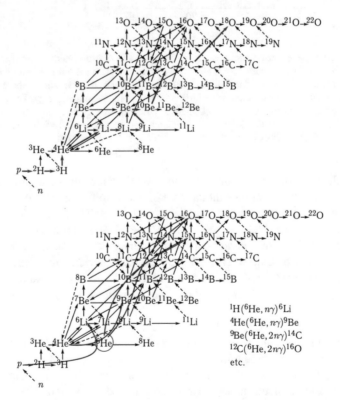

Fig. 6.13 Upper panel: part of the general scheme (network) of astrophysical nucleosynthesis. This comprises the processes of proton and neutron capture, β decay, fusion reactions (radiative capture), etc. Lower panel: the red arrows show possible fusion processes involving the unstable weakly bound ^{6}He nucleus

change the global picture of nucleosynthesis. At low energies the fusion of stable, light nuclei usually occurs as a result of the radiative capture processes (see below), the cross sections of which are small (fractions of a microbarn). The fusion reactions, for example, ^{1}H(^{6}He,^{6}Li+n), ^{3}He(^{6}He,^{7}Be+2n), ^{4}He(^{6}He,^{9}Be+n), ^{9}Be(^{6}He, ^{13}C+2n), etc. should have cross sections of several orders of magnitude greater, since they occur with the participation of nuclear forces. Their inclusion into the general scenario (see the bottom part in Fig. 6.13) can significantly change our understanding of astrophysical nucleosynthesis. Involving such reactions, perhaps, could help to solve the so-called *bottleneck* problem: an unlikely process in carbon synthesis requiring the fusion of three α particles (^{4}He + ^{4}He + ^{4}He → ^{12}C), which must be attracted because of the instability of the ^{8}Be nucleus ($T_{1/2}<10^{-16}$ s). Unfortunately fusion cross sections of nuclei such as ^{6}He with other light nuclei have not yet been measured because sufficiently intense beams of low-energy radioactive nuclei are difficult to obtain.

6.8 Synthesis of Superheavy Elements in Fusion Reactions

The *continent* of stable elements ends with lead ($Z = 82$) and bismuth ($Z = 83$) on the nuclear map (see Fig. 5.15). In nature, however, there are also two heavier elements: thorium ($Z = 90$, $T_{1/2}(^{232}$Th$) = 1.4 \cdot 10^{10}$ years) and uranium ($Z = 92$, $T_{1/2}(^{238}$U$) = 4.5 \cdot 10^9$ years), which were formed by astrophysical nucleosynthesis and have lifetimes comparable with the age of our galaxy. Less stable elements lying between $82 < Z < 90$ are the products of the decay of thorium and uranium nuclei. All known elements with $Z > 92$ were created artificially. At present the elements up to $Z = 118$ have been synthesized; some of their isotopes have lifetimes of several thousand years and can be accumulated in macroscopic amounts (for example, tens of tons of plutonium have been accumulated in the world.) Most artificial elements with $92 < Z \leq 100$ were obtained in nuclear reactors by neutron capture with subsequent β^- decay (plutonium, however, was first isolated when uranium was irradiated with deuterium nuclei in the reaction ^{238}U($d, 2n$)^{238}Np(β^-) → ^{238}Pu). Figure 6.14 shows part of the nuclear map with elements heavier than uranium and a small part of successive neutron capture chains and nuclear decays with formation of increasingly heavier elements.

It is easy to calculate the yield of new elements in neutron-capture processes if we use a system of coupled equations describing the time variation of the number of nuclei of a given isotope, taking into account the probability that the next neutron will be captured by this nucleus and the probability of its decay in a given channel. In a somewhat simplified form, the system of such equations is as follows

$$\frac{dN_{ZA}}{dt} = N_{ZA-1}n_0\sigma^{n\gamma}_{ZA-1} - N_{ZA}n_0\sigma^{n\gamma}_{ZA} - N_{ZA}\lambda^\beta_{ZA} - N_{ZA}\lambda^\alpha_{ZA} - N_{ZA}\lambda^{fis}_{ZA}$$

$$+N_{Z-1A}\lambda^\beta_{Z-1,A} + N_{Z+2A+4}\lambda^\alpha_{Z+2,A+4}. \tag{6.31}$$

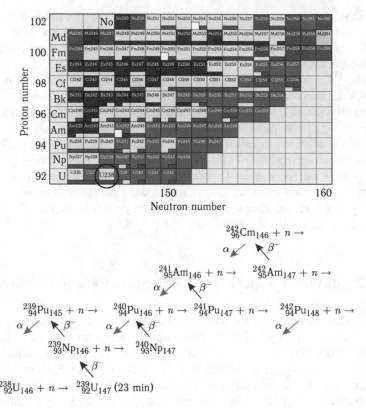

Fig. 6.14 Formation of elements heavier than uranium in nuclear reactors by neutron capture followed by β^- decay

The terms with a minus sign on the right-hand side of the equation describe the decrease in the number of nuclei with given Z and A due to the capture of an additional neutron (transition to $Z, A + 1$), and α decays and fission. The terms with a plus sign correspond to an increase in the number of nuclei Z, A due to neutron capture by the nucleus $Z, A - 1$, due to β^--decay, and due to α decays of the neighboring nuclei. A visual scheme of all these processes is shown in the upper part of Fig. 6.15. The probability of neutron capture by the nucleus Z, A is determined by the neutron flux n_0 (in unit of time per square centimeter), by the neutron-capture cross sections $\sigma_{ZA}^{n\gamma}(E_n)$, and by the decay probability of this nucleus into channel i determined by its half-life $\lambda^i = \ln 2 / T_{1/2}^i$.

At the typical neutron energies produced in the fission of heavy nuclei (several tens or hundreds of keV, that is, essentially nonresonant), the cross sections for neutron capture by atomic nuclei vary fairly smoothly and amount to about 1 barn. Typical neutron fluxes in industrial nuclear reactors generally do not exceed $n_0 = 10^{16}$ neutrons per second per square centimeter, that is, the capture time of one neutron is $\tau_n = \lambda_n^{-1} = 1/(n_0 \sigma_{ZA}^{n\gamma}) \sim 10^8$ s. This time should be compared with the decay-time of the nucleus $Z, A + 1$, formed in such a process. If its lifetime is longer

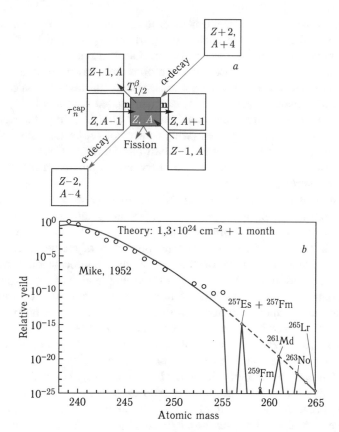

Fig. 6.15 (a) Scheme of formation of new elements in the process of neutron capture and subsequent β^- decay of nuclei in competition with other processes (upper panel). (b) The relative yield of heavy elements found in rock remains after the thermonuclear explosion *Mike* [61], in comparison with calculations performed with the aid of the system of Eq. (6.31)

than τ_n, then it captures the next neutron with formation of a Z, $A+2$ nucleus. This will continue until the lifetime of the next nucleus is much shorter than the time taken to capture the next neutron.

Thus in nuclear reactors it is possible to accumulate only sufficiently long-lived isotopes of heavy elements, to which, in addition, there is access through a chain of successive neutron captures and β^- decays, shown in Fig. 6.14. The heaviest elements that can in principle be obtained in industrial reactors are Einsteinium ($Z = 99$) and Fermium ($Z = 100$). These elements, however, were first obtained not in reactors but in residual fragments of the thermonuclear explosion *Mike*, produced in 1952 (the purpose being far from scientific). Neutron fluxes in such explosions (of the order of 10^{24} neutrons per square centimeter in 1 µs) far exceed neutron fluxes in industrial reactors. The formation of new elements, however, is determined by the same system of Eq. (6.31), the solution of which well agrees with the experimental data shown in Fig. 6.15.

It is impossible to obtain elements heavier than fermium in conventional reactors. If one looks at the map of the nuclei (the part shown in Fig. 6.14), one can see that the isotopes of fermium that can be produced in reactors as a result of the β^- decay of the underlying einsteinium isotopes are short-lived fissioning nuclei. Here the chain to formation of heavier elements through *slow* neutron capture with subsequent β^- decays breaks off. This is the so-called Fermium hole, which can only be bypassed using strong neutron fluxes that allow a shift to the right toward more neutron-rich fermium isotopes (which have not yet been discovered). These should primarily suffer β^- decay rather than fission. Such neutron fluxes can be realized in nuclear explosions (in the laboratory) and in supernova explosions (or in the fusion of neutron stars) in nature.

Today, synthesis of elements heavier than fermium ($Z > 100$) is carried out by fusion reactions using heavy-ion accelerators. Obtaining and studying the properties of new elements (as well as various isotopes) is of great scientific interest, both for nuclear physics and for chemistry. The properties of atomic nuclei can vary considerably with increasing mass and charge. Some theoretical models predict, for example, a sharp decrease in the density in the central region of superheavy nuclei, which is completely uncharacteristic of nuclei already studied. According to the standard liquid-drop model, in nuclei with $Z > 105$ the fission barrier becomes lower than 1 MeV making these nuclei extremely unstable with respect to fission (see Fig. 6.3). However, experiment shows that many of the isotopes of elements with $Z > 105$ have a long lifetime and experience mainly α decay, rather than fission, which indicates that they do in fact have a relatively large fission barrier.

In superheavy nuclei, this fission barrier is exclusively caused by shell effects (that is, the irregular arrangement of single-particle states in the mean field of the nucleus, see Sect. 2.1). Moreover the shell model predicts the existence of "doubly magic nuclei" with filled proton and neutron shells, which should have the greatest binding energy if they lie near to the β stability line. Such nuclei are ^{16}O($Z = 8, N = 8$), ^{40}Ca($Z = 20, N = 20$), and ^{208}Pb($Z = 82, N = 126$) for which the intersection of proton and neutron closed shells occurs precisely on the stability line (see Fig. 5.15).

Theoretical estimates show that with $Z \sim 114$ and $N \sim 184$, the next intersection of the closed proton and neutron shell should occur. The nuclei lying in this region should have an increased stability with respect to fission and α decay. In addition, since these values of Z and N are close to the beta-stability line (see Fig. 5.15), the nuclei from this region (called the *island of stability*) should have a sufficiently long lifetime with respect to β decay. Theoretical models differ significantly in their prediction of the lifetime of these nuclei (from many days to many thousands of years). Discovery of the *island of stability* has been a dream for several generations of physicists. In recent years, it has become possible to significantly approach this island, but it is not yet possible to get onto it (see below). Historically the nuclei that lie exactly on this island were called *superheavy* elements. However, today this term is used to refer to all *trans-Fermium* elements.

The study of the chemical properties of superheavy elements is also of great interest. The order of the elements in the Mendeleyev periodic table is determined by the filling of electron shells whose structure is determined by the nuclear charge. With increasing Z relativistic effects play an increasingly important role in the electronic structure of atoms. These effects are usually calculated using the relativistic Hartree–Fock method for solving the many-body Dirac equation. The effects lead to a change in the radial dependence of the electron wavefunctions for states with different orbital angular momenta and to a possible change in the order of the energy levels due to the increased spin–orbit interaction. Because of this, it is rather difficult to predict the chemical properties of superheavy elements in advance and their experimental measurement is of great interest. Modern methods allow one to make such measurements having just a few atoms with a lifetime of about 1 s or more.

The heaviest target that can be used to produce superheavy nuclei in fusion reactions is californium ($Z = 98$, $T_{1/2}(^{249}\text{Cf}) = 351$ years). If we use accelerated calcium ions ($Z = 20$) as the projectile, then element 118 can be synthesized. To synthesize elements with $Z > 118$ we need beams of even heavier ions. When such heavy nuclei fuse (in contrast to the fusion reactions of lighter nuclei considered above), an important role is played by quasi-fission processes, see Fig. 5.5. It is seen from this figure that in the case of low-energy collisions of ^{16}O and ^{238}U nuclei, in addition to quasi-elastic scattering with the formation of projectile-like and target-like reaction fragments, these nuclei are most likely to fuse to form an excited compound nucleus ^{254}Fm that subsequently fissions into two fragments with masses in the region $A \sim (A_P + A_T)/2$. However, in collisions of heavier ^{48}Ca nuclei with ^{238}U, a much smaller yield of fragments with masses $A \sim (A_P + A_T)/2$ (which could be interpreted as fragments of fission of a compound nucleus) is observed. This indicates a much lower probability of fusion of these nuclei. The reason for this is visible on the same Fig. 5.5 in the mass distribution of all fragments of the reaction $^{48}\text{Ca} + {}^{238}\text{U}$. In addition to possible fission fragments with masses $A \sim (A_P + A_T)/2$ in this reaction an increased yield of fragments with intermediate mass is observed, in the region $A \sim 80$ and $A \sim 200$ in particular.

Thus experimental data on the low-energy collisions of heavy ions (used for the synthesis of superheavy elements) clearly indicate an important role of the quasi-fission process, which significantly reduces the probability of forming a compound nucleus. The entire process of formation of a *cold* superheavy nucleus B (that is, in the ground state), which is the final product of the *cooling* of the excited nucleus C formed in the fusion reaction of two heavy nuclei $A_1 + A_2 \rightarrow C^* \rightarrow B + n, p, \alpha, \gamma$, can be decomposed into three stages, shown schematically in Fig. 6.16.

In the first stage of the reaction the colliding nuclei overcome the Coulomb barrier and come into close contact with their surfaces overlapping. This process competes with elastic and quasi-elastic scattering (including the reactions of few-nucleon transfer, see Chap. 4) with formation of fragments A_1' and A_2', close in mass to the projectile and target. Such competition is highly dependent on the energy of the colliding nuclei and on the impact parameter, that is, on the orbital angular momentum of the relative motion (see the schematic Fig. 5.1). For sub-barrier collision energies, the probability of formation of a contact configuration is small even for zero orbital angular momentum.

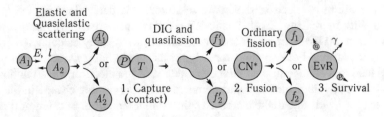

Fig. 6.16 Scenario for superheavy nucleus formation in a fusion reaction

In the second stage of the reaction, the configuration of two touching nuclei should be transformed into a configuration of a more or less spherically symmetric compound nucleus. In the course of this evolution, a heavy system can split into two fragments f_1' and f_2' without compound-nucleus formation. This process is called *quasi-fission*. There are many experimental proofs of a strong competition between the quasi-fission process and the formation of a compound nucleus. The physical reason for this competition becomes understandable if one looks at the potential energy of the system shown in Figs. 5.13 and 5.14. It can be seen from these figures that after formation of the contact configuration of the ^{48}Ca and ^{248}Cm nuclei, it is more advantageous for the system to evolve towards formation of final fragments with masses $A_1 \sim 90$ and $A_2 \sim 200$ (that is, into a channel with mass asymmetry $\eta \sim 0.4$), where the potential energy has a deep minimum caused by a stronger binding energy of nuclei in the region of the doubly magic nucleus $^{208}_{Z=82}Pb_{N=126}$. If, however, a compound nucleus is formed, then it has some excitation energy E^* determined by expression (6.1) and angular momentum ℓ, determined by the impact parameter of the relative motion of the two nuclei for which the fusion occurred. The dividing barriers of heavy nuclei are sufficiently small, and the main channel for the decay of the excited states of these nuclei is fission. If in competition with fission C^* is able to emit several light particles and γ rays that carry away the excitation energy and the angular momentum of this nucleus, then a so-called *evaporation residue* is formed, that is, nucleus B in its ground state. This nucleus is distinguished from the rest of the reaction products by the correlated signals of the detector (located in the focal plane) obtained from the recoil nucleus (having a fixed velocity $\vec{v}_{CN} = A_1\vec{v}_1/(A_1 + A_2)$ and energy) and from the chain of its subsequent α decays and (or) fission.

Thus the cross section for formation of the evaporation residue of the superheavy nucleus in the fusion reaction is still determined by the expression (6.11), which, however, must be supplemented by the probability of formation of a compound nucleus $P_{CN}(A_1 + A_2 \rightarrow C; l, E)$ in competition with quasi-fission (for light and medium-mass nuclei this probability is simply 1)

$$\sigma_{EvR}^{xn,yp,z\alpha}(E) \approx \frac{\pi}{k^2} \sum_{l=0}^{\infty} (2l + 1) P(l, E) P_{CN}(A_1 + A_2 \rightarrow C; l, E)$$

$$\times P_{EvR}^{xn,yp,z\alpha}(C \rightarrow B; l, E^*). \tag{6.32}$$

The same should be done for the fission cross section (6.11) and the total fusion cross section (6.13), which now stands for the cross section for compound-nucleus formation.

Strictly speaking, the formula (6.32) is written in an approximate form, since the entire process of formation of the evaporation residue is broken down into three separate stages of reaction, related to each other but calculated separately. They are: (1) overcoming the Coulomb barrier and forming a contact configuration, (2) forming a composite mono-nucleus in competition with quasi-fission $A_1 + A_2 \rightarrow C^*$, and (3) cooling of the excited compound nucleus in competition with fission $C^* \rightarrow B(\text{g.s.}) + n, p, \alpha, \gamma$. The possibility of such a separation is justified principally by the difference in the time scales of the three stages. The time of passing a distance of several Fermis from the Coulomb barrier to the point of contact of the surfaces of the nuclei does not exceed several units of 10^{-21} s, while the time for the emission of neutrons from the weakly excited compound nucleus is at least two orders of magnitude longer. The intermediate stage of formation of a compound nucleus is not completely independent, being related to both the initial stage of the reaction and to the final stage. In particular, pre-equilibrium emission of light particles is possible at this stage, further complicating its consideration. Nevertheless, the beginning and the end of this stage are well defined in the configuration parameter space, with which the entire reaction is described, and thus the introduction of a separate value P_{CN} to model the intermediate stage of fusion is fully justified in calculating the total cross section for the evaporation-residue formation according to the formula (6.32).

At above-barrier energies, the collision of the heavy nuclei can be analyzed using the Langevin equations (5.5), and the cross sections of the various channels can be calculated from formula (5.9). In this case, there is no need to separate the first and second stages of the fusion reaction. Moreover, it is this approach that makes it possible to accurately follow the competition of fusion and quasi-fission and calculate the value P_{CN} (see below).

It is very tempting to experimentally measure the cross sections of individual processes. If we measure the yield of fragments f_1', f_2', f_1, f_2 and evaporation residues (that is, all reaction products except A_1' and A_2'), we can obtain a cross section for the formation of a contact configuration (or capture cross section) $\sigma_{\text{capt}}(E) = \frac{\pi}{k^2} \sum_{l=0}^{\infty} (2l + 1) P(l, E)$. If in doing so we succeeded in separately obtaining the number of the *true* fission fragments f_1 and f_2, we could obtain the cross section for fusion (formation of a compound nucleus) $\sigma_{CN}(E) = \frac{\pi}{k^2} \sum_{l=0}^{\infty} (2l+1) P(l, E) P_{CN}(l, E)$. Since the statistical model gives sufficiently reliable predictions of the decay probabilities of an excited compound nucleus through various channels (see Sect. 6.2), then as a result of such measurements it might be possible to make a reliable estimate of the cross section for superheavy-nucleus formation (6.32) for a given combination of nuclei. Such estimates are crucial because of the complexity and high cost of carrying out experiments on the synthesis of superheavy elements.

However, it is rather difficult (sometimes, impossible) to experimentally distinguish the *true* fission fragments f_1 and f_2. In the process of quasi-fission, fragments f_1' and f_2' can also be formed and they are close in mass to the fragments of true fission of the compound nucleus f_1 and f_2 (see the trajectory of QF$_2$ in Figs. 5.13 and 5.14). For mass-symmetric reactions, that is, in the case $A_1 \sim A_2$ at low collision energies, it is absolutely impossible to distinguish the fragments A_1', A_2', f_1', f_2', f_1, f_2 formed at different stages of the reaction.

The first stage of formation of a superheavy nucleus, the process of nuclear capture, is the most easy to understand. For any combination of colliding nuclei the corresponding cross section

$$\sigma_{\text{capt}}(E) = \frac{\pi}{k^2} \sum_{l=0}^{\infty} (2l+1) P(l, E)$$

is simple enough to predict if their static deformations and their vibrational properties are known (see Sect. 6.4). To calculate the nucleus–nucleus potential a folding procedure or a proximity potential can be used (see Sect. 2.2). The accuracy of prediction of the capture cross section in the most interesting near-barrier energy region is better than a factor 2, which is perfectly acceptable for planning an experiment.

An estimate of the *survivability* of the excited compound nucleus and a calculation of the yield of the final nucleus $B(\text{g.s.}) = C^* - xn$ in the channel with the evaporation of x neutrons (for superheavy nuclei only neutron evaporation competes with fission, see Fig. 6.5) can be made within the framework of the statistical model. This model and its parameters are well developed on numerous experimental data. The description of this model is given above in Sect. 6.2, and formula (6.9) determines the probability of survival of the compound nucleus in the channel with the evaporation of x neutrons, $P_{\text{EvR}}^{xn}(C \to B + xn)$, which is included in formula (6.32).

Figure 6.17a,b shows the cross sections for formation of the californium isotopes ^{243,244,245}Cf ($Z = 98$) in 3n, 4n, and 5n evaporation channels and the isotope of hassium ^{265}Hs ($Z = 108$) in the fusion reactions ^{12}C+^{238}U [61] and ^{58}Fe + ^{208}Pb [33], respectively. The same figure also shows the capture cross sections for these reactions measured in [48] and [35]. The theoretical cross sections for the yields of evaporation residues were obtained using formula (6.32) under the assumption that there is no quasi-fission at the formation stage of the compound nucleus, that is, $P_{\text{CN}} = 1$. It can be seen that this assumption is justified by the fusion of sufficiently light ^{12}C nuclei with uranium nuclei. However, in the case of the fusion of the heavier ^{58}Fe projectile nucleus with ^{208}Pb, the probability of compound-nucleus formation P_{CN} is less than 1%, and neglecting this leads to an overestimated evaporation-residue cross section by more than two orders of magnitude (see Fig. 6.17b).

The cross section for the yield of this specific hassium evaporation residue has a characteristic bell-shaped appearance with a width of the order of 5 MeV at half-

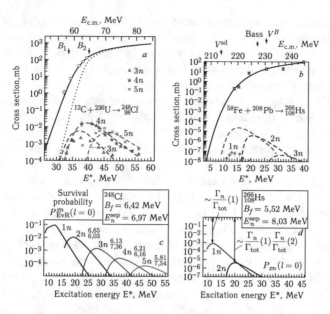

Fig. 6.17 Cross sections for capture and the yield of evaporation residues in the fusion reactions ^{12}C + ^{238}U (**a**) and ^{58}Fe + ^{208}Pb (**b**). The experimental data were taken from [61] and [48] for the first reaction and from [33] and [35] for the second one. The dotted and solid curves show the calculated results of the capture cross section in the penetrability model of a one-dimensional barrier, and when taking account of the channel coupling (see Sect. 6.4). The cross sections for the evaporation residues were calculated using formula (6.32) under the assumption that there is no quasi-fission at the stage of formation of the compound nucleus, that is, $P_{CN} = 1$. Figures (**c**) and (**d**) show the probabilities of survival in the channels with emission of x neutrons, calculated from formula (6.9)

maximum. This is simply explained by the fact that as the collision energy decreases the fusion cross section rapidly decreases in the sub-barrier region, and as the collision energy increases the excitation energy of the compound nucleus increases reducing the survivability (see the lower panels in Fig. 6.17c,d). Thus the maximum yield of evaporation residues is expected at collision energies close to the Coulomb barrier. Since the cross sections for superheavy-nucleus formation are extremely small (of the order of 1 pb), a careful choice of collision energy is very important: at a cross section of 1 pb a one-week irradiation of the target is required to observe just one event. A 5 MeV error in the choice of collision energy can lead to a nul result from an extremely expensive experiment.

If the collision energy is chosen close to the height of the Coulomb barrier V^B, then the excitation energy of the resulting compound nucleus depends on the combination of the colliding nuclei, see formula (6.1). For mass-asymmetric combinations (for example, ^{12}C + ^{238}U) the excitation energy at $E_{c.m.} \sim V^B$ is about 40 MeV, and for more symmetric combinations involving strongly bound nuclei (for example, ^{58}Fe + ^{208}Pb) this energy is much lower (see Fig. 6.17). Reactions of the first type (with a high excitation energy) are usually called *hot* synthesis, and fusion reactions of the second type are called *cold* synthesis (based on the temperature of the compound nucleus formed). The emission of one neutron reduces the excitation energy of the nucleus by about 10 MeV (about 8 MeV of neutron binding energy plus around 2 MeV of kinetic energy; see Fig. 6.5a). Thus in order to remove the 40 MeV excitation energy, the nucleus must evaporate four neutrons. This is why in *hot*-synthesis reactions the maximum evaporation-residue cross section occurs in channels with the emission of three–five neutrons, and for *cold* synthesis, one–two neutrons.

In principle, an evaporated neutron can have any kinetic energy in the range $0 < e < E^*$. However, the probability that a neutron will carry away a large kinetic energy falls off exponentially with growing e, and is determined by the factor $P_n(E^*, e) = C\sqrt{e}\exp[-e/T(E^*)]$ in formula (6.9) for the probability of formation of a cold, final nucleus in a channel with the emission of x neutrons. The Maxwellian velocity distribution of evaporated neutrons (shown schematically in Fig. 6.5) explains the smooth exponential decrease in the values of P_{EvR}^{xn} in Figs. 6.6 and 6.17 with increasing excitation energy of the compound nucleus, as well as the overlapping of the different evaporation channels as a function of excitation energy.

Two major factors make the cross sections for the formation of superheavy elements very small: a low probability P_{CN} of compound nucleus formation in competition with the dominant quasi-fission process, and a low probability of survival of the excited compound nucleus P_{EvR}^{xn}. The first factor depends on the combination of colliding nuclei and also on the collision energy (see below), whereas the survival of the compound nucleus depends on its excitation energy and on the properties of this nucleus (fission barrier and neutron separation energy). The probability of neutron emission (which removes part of the excitation energy of the compound nucleus) is determined by the ratio of the widths $\Gamma_n/\Gamma_{\text{tot}}$ (see Sect. 6.2). For excited superheavy nuclei the main decay channel is fission (see Fig. 6.6) and $\Gamma_{\text{tot}} \approx \Gamma_f$. From the formulas (6.2)–(6.4) for the decay widths and formula (6.7) for the level density, one can obtain a rough estimate of the probability of neutron emission as a function of its separation energy E_n^{sep} and the height of the fission barrier B_{fis}:

$$\Gamma_n/\Gamma_f \sim \exp\left[-\alpha(E_n^{\text{sep}} - B_{\text{fis}})/T(E^*)\right],$$

where $T(E^*) = \sqrt{E^*/a}$ is the temperature of the compound nucleus. Thus, the smaller the fission barrier, and the greater the neutron binding energy, the less is the probability of its survival. For example, we can compare the survival probabilities (at the same excitation energy) of ^{248}Cf ($E_n^{\text{sep}} - B_{\text{fis}} \sim 0.5$ MeV) and ^{266}Hs ($E_n^{\text{sep}} - B_{\text{fis}} \sim 2.5$ MeV) shown in the lower part of Fig. 6.17. It is evident that for the second nucleus, which has a smaller fission barrier height and a large neutron binding energy, the survival probability is much leower.

As noted in Sect. 6.2, the macroscopic component of the fission barrier for superheavy nuclei is very small and their stability with respect to fission is almost completely determined by quantum shell effects. However, the microscopic component of the fission barrier, δW, caused by these effects (see above) decreases with increasing compound-nucleus excitation energy (the shell effects are damped out with increasing E^*), which further reduces the probability of its survival.

The process of overcoming the Coulomb barrier in the entrance channel (followed by contact of the surfaces of the colliding nuclei), as well as the process of *cooling down* the excited compound nucleus by evaporation of light particles that occur in competition with the dominant fission channel, are quite clear. The corresponding probabilities $P(l, E)$ and $P_{EvR}^{xn,yp,z\alpha}(C \rightarrow B + xn + yp + z\alpha)$ in formula (6.32) are calculated fairly accurately. However, the modeling and calculation of the corresponding probability P_{CN} of the intermediate process of compound-nucleus formation from the configuration of two touching nuclei is much more complicated. Microscopic analyses of the collision dynamics of two heavy nuclei (performed using the time-dependent Schrodinger equation or the time-dependent Hartree–Fock method) show that after the surfaces come into contact there is a slow (adiabatic) change of the shape from the touching configuration into a more or less spherical compound nucleus. In this case, the mean fields of these nuclei are gradually united. The high velocity of nucleons (in comparison with the velocity of approach of the nuclei) allows them to *adjust* their motion in the changing mean field while maintaining the volume and density of the nuclear matter. The dependence of the potential energy of the system on the parameters determining the specific nuclear configuration (the distance between the centers, the mass and charge asymmetry, deformations), as well as the single-particle states, which determine the shell correction to this energy, can be calculated within the framework of a two-center shell model (see Sect. 5.2).

At above-barrier energies the solution of the dynamical equations (5.5) allows one to trace the entire collision process, including the evolution of the nuclear system from the configuration of two touching nuclei to the compound-nucleus configuration. Along the way the quasi-fission of the nuclear system is the most probable decay mode (see Figs. 5.13 and 5.14). However, a small fraction of the trajectories do attain the configuration of a more or less spherically symmetric compound nucleus. The ratio of the number of such trajectories to the total number of events tested (for a given impact parameter) determines the probability of formation of a compound nucleus $P_{CN}(l, E)$.

The cross sections of capture, fusion (formation of the compound nucleus), and the evaporation residues of the superheavy nuclei formed in *cold*-synthesis reactions using a ^{208}Pb target are shown in Fig. 6.18. The same figure shows the dependence of the probability for compound nucleus formation on its excitation energy and on the combination of colliding nuclei. This dependence, shown in Fig. 6.18b, can be approximated by the simple expression

$$P_{CN}(l, E*) = \frac{P_{CN}^0}{1 + \exp\left[\frac{E_{Bass}^* - E_{int}^*(l)}{\Delta}\right]}, \tag{6.33}$$

which can be used for a quick estimate of the cross section for the yield of superheavy nuclei in the reactions of *cold* fusion. In this expression E_{Bass}^* is the

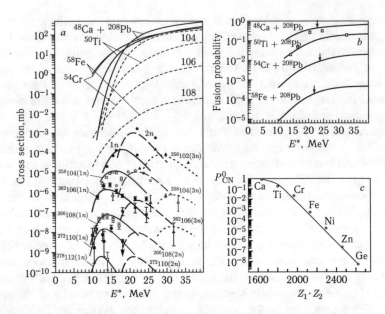

Fig. 6.18 (**a**) The capture cross sections (upper solid curves), fusion cross sections (short dashed curves), and the evaporation-residue cross sections (formation of superheavy nuclei) in the reactions of *cold* fusion for various projectiles (from ^{48}Ca to ^{70}Zn) on a ^{208}Pb target. The experimental data were taken from [33] and [46] (for element 110). (**b**) Calculated values of the probability of compound nucleus formation $P_{CN}(l = 0, E)$ in the same heavy-ion fusion reactions with ^{208}Pb nuclei. The experimental data for ^{50}Ti + ^{208}Pb were taken from [50]. (**c**) The above-barrier probability of nuclear fusion. The points are numerical calculations, the curve corresponds to the formula (6.34)

excitation energy of the compound nucleus obtained at a collision energy equal to the height of the Coulomb barrier (calculated, for example, with the Bass potential), that is, $E^*_{\text{Bass}} = V^B_{\text{Bass}} + Q_{\text{fus}}$ (see formula (6.1) and Sect. 2.2.4). The values of this energy are shown by the arrows in Fig. 6.18b. Also in formula (6.33) $E^*_{\text{int}}(l) = E_{\text{c.m.}} + Q_{\text{fus}} - E^{\text{rot}}_{\text{CN}}(l)$ is the *internal* excitation energy of the compound nucleus ignoring its energy of rotation $E^{\text{rot}}_{\text{CN}} = (\hbar^2/2\mathcal{J}_{\text{g.s.}})l(l + 1)$, Δ is an adjustable parameter approximately equal to 4 MeV, and P^0_{CN} is an *asymptotic* (above-barrier) fusion probability, depending only on the combination of colliding nuclei. The P^0_{CN} values calculated for an excitation energy of 40 MeV (that is, much higher than E^*_{Bass}) are shown by the dots in Fig. 6.18c for various combinations of colliding nuclei. It is obvious that the probability of fusion of heavy nuclei decreases exponentially with increasing $Z_1 \cdot Z_2$. The dependence of P^0_{CN} on $Z_1 \cdot Z_2$ can be approximated by the empirical formula

$$P^0_{\text{CN}} = \frac{1}{1 + \exp\left[\dfrac{Z_1 Z_2 - \zeta}{\tau}\right]}, \qquad (6.34)$$

shown by the solid curve in Fig. 6.18c. Here, $\zeta \approx 1760$ and $\tau \approx 45$ are simply adjustable parameters.

The approximate expressions (6.33) and (6.34) for the probability of compound nucleus formation are applicable only for *cold*-fusion reactions involving the doubly magic nucleus ^{208}Pb or the nearby ^{209}Bi. For reactions of *hot* synthesis using heavier actinide targets, sufficiently simple empirical formulas for $P_{CN}(l, E)$ have not yet been found. To make any estimates of the cross sections for the formation of superheavy elements in these reactions, it is necessary to carry out rather complicated calculations, for example, with the aid of Eq. (5.5). In addition to their complexity such calculations are not entirely reliable due to uncertainties in the values of some of the fundamental nuclear-dynamics quantities, for example, the transfer rate of nucleons λ_0 or the magnitude of the nuclear viscosity (see Sect. 5.3). In spite of this, predictions of cross sections for the formation of superheavy elements in ^{48}Ca fusion reactions with actinide targets, made within the framework of the model described above, are in good agreement with experimental data (see Fig. 6.19).

It was in the fusion reactions of accelerated ^{48}Ca ions at the turn of the twenty first century that six new elements were synthesized in the JINR laboratory (Dubna, Russian Federation); this started with $Z = 113$ and ended with the element $Z = 118$ obtained in the ^{48}Ca ($Z = 20$) fusion reaction with the heaviest target of ^{249}Cf ($Z = 98$). Isotopes of the elements with $Z > 98$ are obtained in nuclear reactors in microscopic amounts that are insufficient for preparation of a usable target. Consequently, to obtain superheavy elements with $Z > 118$ we must use projectiles heavier than ^{48}Ca. In this case the probability of fusion significantly decreases. Figure 6.20 shows the predicted cross sections for the formation of elements 119

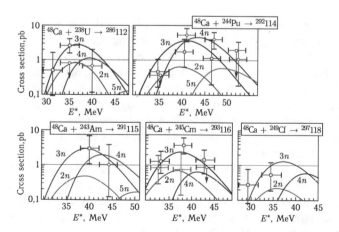

Fig. 6.19 Cross sections for superheavy-nucleus formation in fusion reactions of ^{48}Ca with various actinide targets (from uranium to californium) in the 2n (triangles), 3n (squares), 4n (circles), and 5n evaporation channels. The experimental data were taken from the review paper [52], and the theoretical predictions were made within the framework of the approach described above

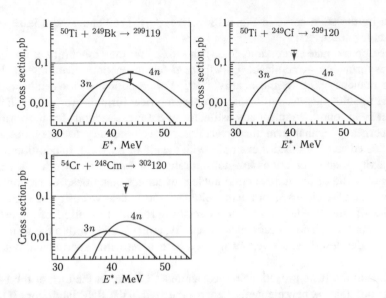

Fig. 6.20 The predicted cross sections for the creation of the superheavy elements 119 and 120 in the fusion reactions ^{50}Ti $+$ ^{249}Bk, ^{50}Ti $+$ ^{249}Cf, and ^{54}Cr $+$ ^{248}Cm. The arrows show the upper limits of the cross sections achieved in the GSI [32, 39]

and 120 in the reactions ^{50}Ti $+$ ^{249}Bk, ^{50}Ti $+$ ^{249}Cf, and ^{54}Cr $+$ ^{248}Cm, and the upper limits of the experimental cross sections obtained at the GSI (Darmstadt, Germany). So far (2015) these elements have not yet been synthesized, however, relevant experiments are on-going in Dubna, in the GSI and in the RIKEN laboratory (Japan).

As noted above, because of the increasing slope of the stability line towards the neutron axis (increasing N/Z ratio with increasing A) only proton-rich isotopes of heavy elements, lying to the left of the stability line, can be formed in fusion reactions with stable nuclei (see Fig. 5.15). Recently it has become possible to obtain and accelerate short-lived radioactive nuclei, including neutron-rich nuclei (see below). In principle these could be used to synthesize neutron-rich isotopes of superheavy elements located in the center of the *island of stability*. For example, in the fusion reaction ^{44}S($T_{1/2}$=0.1 s) $+$ ^{248}Cm, it might be possible to obtain isotopes of element 112, having six neutrons more than the isotopes obtained in the fusion of stable ^{48}Ca $+$ ^{238}U nuclei. However, the cross sections for the formation of these isotopes in this reaction are also extremely small (of the order of 1 pb) and for their production a ^{44}S ion beam would be required with the same intensity as the ^{48}Ca ion beam, that is, about 10^{12} particles per second. Unfortunately such intense beams of radioactive nuclei have not yet been obtained.

Figure 6.21 shows a modern (2014) chart of the nuclei in the region of the superheavy elements. In the near future it is planned to synthesize elements 119 and 120 in the fusion reactions ^{50}Ti+^{249}Bk, ^{50}Ti+^{249}Cf, and/or ^{54}Cr+^{248}Cm. The isotopes of these elements will again be located in the region of proton-rich nuclei,

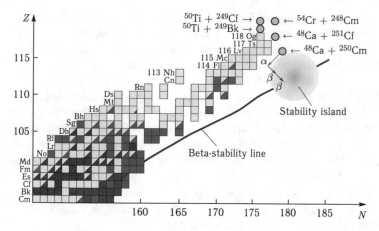

Fig. 6.21 The top of the nuclear chart. The already synthesized isotopes and their dominant decay modes have been marked. The circles denote isotopes that are planned to be obtained in the near future in the fusion reactions indicated in the same figure

far from the island of stability. The existence of this island is indirectly proved by an increase in the lifetime of the known isotopes of 112 and 113 elements by several orders of magnitude as they approach the island: $T_{1/2}(^{277}112) = 0.7\,\text{ms}$, $T_{1/2}(^{285}112) = 30\,\text{s}$, $T_{1/2}(^{278}113) = 0.24\,\text{ms}$, $T_{1/2}(^{286}113) = 13\,\text{s}$. The hypothetical possibility of getting to the island of stability with the help of fusion reactions can emerge if for some isotopes of 114 and/or 115 elements the probability of β^+ decay proves commensurate with the probability of their α decay or spontaneous fission (which can be predicted in some theoretical models). For example, an isotope $^{295}117$ is formed in the 2n channel of the $^{48}\text{Ca}+^{249}\text{Bk}$ fusion reaction (cross section 0.2 pb). After α decay this turns into the isotope $^{291}115$, which can undergo three successive β^+ decays to form the superheavy nucleus $^{291}112$ whose lifetime is estimated to be 1000 years. The same possibility also appears in the 3n evaporation channel of the $^{48}\text{Ca}+^{250}\text{Cm}$ fusion reaction (with a cross section of 0.8 pb), shown in Fig. 6.21.

6.9   Radiative Capture of Light Nuclei

The fusion of light nuclei is energetically favorable, that is, $Q_{\text{fus}} = E_{\text{bind}}(A_{\text{CN}}) - E_{\text{bind}}(A_1) - E_{\text{bind}}(A_2) > 0$ (see Formula (6.1)). This means that such a reaction could occur even close to zero kinetic energy, if not for the obstacle of the Coulomb barrier which the nuclei must overcome to form a more bound final nucleus. Such fusion reactions of light nuclei are one of the processes responsible for nucleosynthesis and energy release in stars.

However, the mechanism for forming a heavier nucleus by the coalescence of light nuclei differs sharply from the mechanism of fusion of heavy systems. At low collision energies there may simply not be a suitable state (energy level) in the final nucleus that will ensure energy conservation $E_{c.m.} + m_1c^2 + m_2c^2 = m_{CN}c^2 + \varepsilon_\nu(CN)$. This means that in the fusion of light nuclei, a *third* particle must participate in order to carry away the excess energy. If the excitation energy of the final nucleus is lower than the neutron separation energy, then the only possibility for this third particle is a γ ray. A typical example of the fusion of light nuclei is shown in Fig. 6.22. The mass of the fusing ^{3}H and ^{4}He nuclei exceeds the mass of the ^{7}Li nucleus by 2468 keV. At low collision energies there is no corresponding excited state in the ^{7}Li nucleus that can be formed by the fusion. This means that when a ^{7}Li nucleus is formed, a γ ray with energy $E_{c.m.} + 2468$ keV (with the formation of ^{7}Li in the ground state) or two γ rays with energies $E_1 = E_{c.m.} + 2468$ keV and $E_2 = 478$ keV must be emitted (see Fig. 6.22).

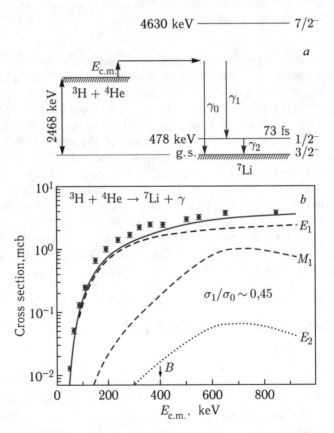

Fig. 6.22 Scheme (**a**) and cross section (**b**) of the radiative capture (fusion) of ^{3}H and ^{4}He nuclei with formation of the ^{7}Li nucleus. The experimental data were taken from [15]. The ^{7}Li production cross sections in the ground state with emission of $E1$, $E2$, and $M1$ γ-rays were calculated within the potential model of radiative capture on the web site http://nrv.jinr.ru

Such reactions are called radiative-capture processes. Since they occur via the electromagnetic interaction and at very small distances compared to the wavelength of the photon (limited by the volume of the nucleus), their cross sections are much smaller than the fusion cross sections discussed above. Even at above-barrier energies the cross section for the radiative capture of ^{3}H and ^{4}He nuclei does not exceed 10 μb (see Fig. 6.22) which is five orders of magnitude smaller than the ^{4}He fusion cross section with heavy nuclei (see, for example, Fig. 6.12). The cross sections for radiative capture are generally measured from the yield of γ rays. For the reaction ^{3}H(^{4}He,γ)^{7}Li, in which there are two possibilities for the formation of a final nucleus (direct formation of the ground state with emission of γ_0, as well as the population of the excited state with emission of γ_1 and its subsequent decay to the ground state with emission of γ_2), one can also measure the branching factor $R = \sigma_1/\sigma_0$. In this reaction a value of $R \approx 0.4$ was found which turned out to be practically independent of the collision energy.

The cross section for radiative capture is determined by the overlap of the wavefunctions of the initial and final states of the nuclear system and by the electromagnetic interaction leading to the production of a photon. The energy of the photon and its momentum are determined by energy conservation $E_\gamma = \hbar c k_\gamma = E - E_f$. Here, E is the collision energy in the center-of-mass system, and $E_f = m_{CN}c^2 - (m_1c^2 + m_2c^2)$ is the (negative) binding energy of the particles trapped in the final nucleus (= -2468 keV for this reaction). If the final nucleus is formed in some excited state, then we must add the corresponding excitation energy to E_f. Since $\hbar c = 197$ MeV $\times$ fm, the typical values of the photon momentum are $k_\gamma \sim 0.05$ fm^{-1} at $E_\gamma \sim 1$ MeV. This means that at the nuclear distances where radiative capture takes place, $k_\gamma r \ll 1$, and in the multipole expansion of the interaction of the electromagnetic field with the charged system, only the lower multipoles make a significant contribution since the Bessel functions in this expansion rapidly decrease with increasing λ: $j_\lambda(kr) \approx (kr)^\lambda/(2\lambda + 1)!!$ (where $\lambda = 1, 2, \ldots$ is the multipolarity of the electric or magnetic transition).

The cross section for radiative capture with formation of a final nucleus with angular momentum J_f and parity π_f is (in a somewhat simplified form)

$$\sigma_\gamma^{J_f \pi_f}(E) = \sum_{\lambda \geq 1, l \geq 0, J_i} C(J_f, \lambda, l, J_i) k_\gamma^{2\lambda+1} \left| \left\langle \varphi_{CN}^{J_f \pi_f} \left\| \mu_\lambda^E + \mu_\lambda^M \right\| \psi_l^{J_i \pi_i}(E) \right\rangle \right|^2 .$$

$$(6.35)$$

Here, $\varphi_{CN}^{J_f \pi_f}(r)$ is the wavefunction of the final nucleus describing the bound state of the fused nuclei (taking into account the corresponding spectroscopic factor), $\varphi_{CN}^{J_f \pi_f}(r \to \infty) \sim \exp(-\kappa_f r)$, $\kappa_f = \sqrt{(2\mu_{12}/\hbar^2)|E_f|}$, and $\psi_l^{J_i \pi_i}(E; r)$ is the partial wavefunction of the relative motion of the colliding nuclei in the entrance channel. This function satisfies the radial Schrödinger equation (3.15) with the correctly chosen interaction between the nuclei A_1 and A_2. Since the multipole operator of the electric transition behaves as $\mu_\lambda^E \sim er^\lambda$ and that of the magnetic transition as $\mu_\lambda^M \sim \frac{e\hbar}{2mc} r^{\lambda-1}$, then magnetic transitions of the same multipolarity are on average less likely than electric transitions by a factor $(\hbar/mcR_{\text{nuc}})^2$ (m is the nucleon mass), that is, by approximately two orders of magnitude (this is only a rough estimate, since the actual cross section also depends on the wavefunctions of the initial and final states). The transition probability also decreases with increasing multipolarity of the emitted γ ray. When λ is increased by one, the transition probability decreases by the factor $(k_\gamma R_{\text{nuc}})^2/(2\lambda+3)^2 \ll 1$.

From angular momentum conservation $\vec{J}_i = \vec{s}_1 + \vec{s}_2 + \vec{l}$ ($\vec{s}_1$ and $\vec{s}_2$ are the spins of the colliding nuclei) and $\vec{J}_f + \vec{\lambda} = \vec{J}_i$. In this process parity is also conserved but it should be remembered that the parity of an electric transition is equal to $(-1)^\lambda$ and that of a magnetic transition is $(-1)^{\lambda+1}$. These conservation laws (together with the above-mentioned suppression factors) lead to a substantial reduction in the number of possible transitions in near-barrier radiative-capture processes. As a rule, the main contributions are made by $E1$, $E2$, and $M1$ transitions and by lower partial waves.

In the potential model of radiative capture it is the relative motion of the merging particles (and not their internal structure) that is assumed to play the main role. To describe this motion in the entrance channel and in the newly formed nucleus, it is necessary to select the corresponding interaction potentials, which, in principle, can differ from each other. The potential energy describing the bound state of the nuclei A_1 and A_2, should reproduce the correct binding energy E_f and the required quantum numbers of this state. The potential for the interaction of nuclei A_1 and A_2 should also properly describe their elastic scattering at low energies. In particular, if this potential is correctly selected, the resonances observed in the experiment (if any) should also be reproduced. From the theory of elastic scattering it is known that when passing through (in energy) such a potential resonance, the corresponding partial phase-shift $\chi_l(E)$ (see Chap. 3) must drastically change (approximately by π). This behavior of the partial scattering phase helps to select the appropriate interaction potential of the nuclei in the entrance channel in the correct way.

An example of resonant radiative capture of protons by carbon nuclei is shown in Fig. 6.23. From the level diagram of the excited states of the compound nucleus ^{13}N, one can see that at an energy of relative motion of the ^{1}H and the ^{12}C equal to 421 keV, the resonance should be observed for the partial wave $\ell = 0$. The correct position of this resonance is obtained by choosing the ^{1}H-^{12}C interaction in Woods–Saxon form with the parameters $V_0 = -63.45$ MeV, $r_0^V = 1.179$ fm, and $a_V = 0.65$ fm. Their $1p_{1/2}$ bound state with binding energy $E_f = -1.944$ MeV is reproduced by a potential with a depth of $V_0 = -54.33$ MeV and with the same radius. The cross section for the radiative capture of ^{1}H by ^{12}C calculated with these potentials coincides extremely well with experimental data [41, 55].

The most interesting fusion processes of light nuclei are those at low (sub-barrier) energies of the order of several keV or tens of keV. The temperature inside the Sun, for example, is approximately 1.5×10^7 K, which corresponds to $kT \approx 1.3$ keV. The energy distribution of the colliding nuclei is determined by the Boltzmann law $\exp(-E/kT)$, that is, it decreases exponentially with increasing E. At such low energies it is very difficult to carry out experimental measurements of the

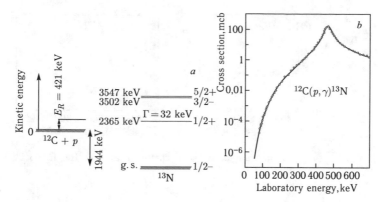

Fig. 6.23 Scheme (**a**) and cross section (**b**) of the radiative capture of ^{1}H and ^{12}C nuclei to form the ^{13}N nucleus. The experimental data were taken from [41, 55]. The capture cross section was calculated in the framework of the potential model on the web site http://nrv.jinr.ru

corresponding cross sections because of their smallness, and a simple extrapolation of the cross sections to the low-energy region is difficult because of their sharp (exponential) decrease in this region (see Figs. 6.22 and 6.23). It is quite obvious that such a reduction in the cross section in the sub-barrier energy region is exclusively due to the repulsive Coulomb force which prevents the nuclei from approaching to a distance $r = R_1 + R_2$, where fusion can occur. It is easy to calculate that at a collision energy $E = 100$ keV the ^{1}H and ^{12}C can approach only to a distance $r_0 = Z_1 Z_2 e^2 / E = 1 \times 6 \times 1.44/0.1 \approx 86$ fm, which is much larger than the sum of the radii of these nuclei. Their further approach is possible only by quantum tunneling.

The behavior of the wavefunction of relative motion $\psi_l(E; r)$ at $r > R_1 + R_2$ is completely determined by the Coulomb field $V_C(r) = Z_1 Z_2 e^2 / r$. Since this function must have a finite value for all r, then at small distances $\psi_l(E; r) \sim F_l(E, r)$, where $F_l(E; r)$ is the regular Coulomb wavefunction. At small distances this function (for $kr \ll \eta$) behaves as follows: $F_{l=0}(E; r) \sim \frac{1}{2} e^{-\pi \eta} \left(\frac{kr}{2\eta} \right)^{1/4}$ [1] (it is the partial wave $\ell = 0$ that makes the main contribution to the cross section at low energies). Here $\eta = k(Z_1 Z_2 e^2)/2E$ is the Coulomb parameter which increases with decreasing collision energy. Thus, the probability that the nuclei should come into contact (and the process of fusion occur) at low (sub-barrier) collision energies decreases as $e^{-2\pi \eta}$ (that is, as $e^{-B/\sqrt{E}}$, where $B = \pi \sqrt{2\mu} Z_1 Z_2 e^2 / \hbar$, and μ is the reduced mass of the colliding nuclei).

We can single out this apparent energy dependence of the fusion cross section (not connected with the subsequent nuclear process), as well as the kinematic factor $1/E$ in Eq. (6.13), and write down the fusion cross section (or other sub-barrier nuclear reaction) as

$$\sigma_{fi}(E) = \frac{1}{E} e^{-2\pi \eta} S_{fi}(E). \qquad (6.36)$$

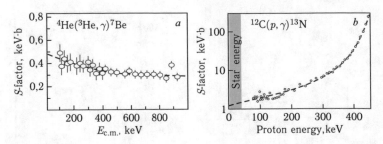

Fig. 6.24 Astrophysical S-factors for the radiative capture reactions $^4\mathrm{He}(^3\mathrm{He},\gamma)^7\mathrm{Be}$ **(a)** and $^{12}\mathrm{C}(\mathrm{p},\gamma)^{13}\mathrm{N}$ **(b)**

The quantity $S(E)$ is called the astrophysical S-factor. It is this value that determines the probability of a particular process occurring when the colliding nuclei come in contact. The value of the S-factor depends on the initial and final states of the nuclei and on the interaction that causes the given reaction, for example, electromagnetic in the case of radiative capture or the weak interaction in hydrogen burning in stars: $p + p \rightarrow d + e^+ + \nu$. The dependence of the S-factor on energy should not be very strong, allowing one to extrapolate its value to very low energies, where it is impossible to carry out the corresponding experiment. It is this possibility that motivates the introduction of the S-factor.

Figure 6.24 shows the energy dependence of the S-factors for the radiative capture reactions $^4\mathrm{He}(^3\mathrm{He},\gamma)^7\mathrm{Be}$ and $^{12}\mathrm{C}(\mathrm{p},\gamma)^{13}\mathrm{N}$. The sharp growth in the $^{12}\mathrm{C}(\mathrm{p},\gamma)^{13}\mathrm{N}$ S-factor at higher energy is due to its resonant character at 421 keV (see above). S-factors of various astrophysical processes can be found in the Internet database, see for example www.astro.ulb.ac.be/nacreii/.

References

1. M. Abramovitz, I. Stegun, *Handbook of Special Functions: with Formulas, Graphs, and Mathematical Tables* (Dover Publications, New York, 1974)
2. G.D. Alkhazov, Izv. AN SSSR, Ser. Fiz. **41**, 21 (1977)
3. G.D. Alkhazov, T. Bauer, R. Beurtey, A. Boudard, G. Bruge, A. Chaumeaux, P. Couvert, G. Cvijanovich, H.H. Duhm, J.M. Fontaine, D. Garreta, A.V. Kulikov, D. Legrand, J.C. Lugol, J. Saudinos, J. Thirion, A.A. Vorobyov, Elastic and inelastic scattering of 1.044 GeV protons by ^{40}Ca, ^{42}Ca, ^{44}Ca, ^{48}Ca and ^{48}Ti. Nucl. Phys. A **274**(3), 443–462 (1976)
4. N. Anantaraman, Systematic study of the ^{88}Sr(^{16}O, ^{15}N) reaction. Phys. Rev. C **8**, 2245–2254 (1973)
5. D. Bachelier, M. Bernas, J.L. Boyard, H.L. Harney, J.C. Jourdain, P. Radvanyi, M. Roy-Stephan, R. Devries, Exchange effect in the 166 MeV α-particle elastic scattering on ^{6}Li. Nucl. Phys. A **195**(2), 361–368 (1972)
6. A.R. Barnett, J.S. Lilley, Interaction of alpha particles in the lead region near the coulomb barrier. Phys. Rev. C **9**, 2010–2027 (1974)
7. R. Bass, *Nuclear Reactions with Heavy Ions* (Springer, Berlin, 1980)
8. A.M. Bastawros, C.L. Bennett, H.W. Fulbright, R.G. Markham, The ^{64}Ni(^{6}Li, d)^{68}Zn reaction. Nucl. Phys. A **315**(3), 493–499 (1979)
9. F.D. Becchetti, G.W. Greenlees, Nucleon-nucleus optical-model parameters, A> 40, E< 50 MeV. Phys. Rev. **182**, 1190–1209 (1969)
10. S. Beghini, C. Signorini, S. Lunardi, M. Morando, G. Fortuna, A.M. Stefanini, W. Meczynski, R. Pengo, An electrostatic beam separator for evaporation residue detection. Nucl. Instrum. Methods Phys. Res. Sect. A **239**(3), 585–591 (1985)
11. M. Bender, P.-H. Heenen, P.-G. Reinhard, Self-consistent mean-field models for nuclear structure. Rev. Mod. Phys. **75**, 121–180 (2003)
12. H.G. Bingham, M.L. Halbert, D.C. Hensley, E. Newman, K.W. Kemper, L.A. Charlton, Mirror states in $A = 15$ from 60 MeV ^{6}Li-induced reactions on ^{12}C. Phys. Rev. C **11**, 1913–1924 (1975)
13. J. Blocki et al., Proximity forces. Ann. Phys. **105**, 427–462 (1977)
14. H.G. Bohlen, E. Stiliaris, B. Gebauer, W. von Oertzen, M. Wilpert, T. Wilpert, A. Ostrowski, D.T. Khoa, A.S. Demyanova, A.A. Ogloblin, Refractive scattering and reactions, comparison of two systems: ^{16}O + ^{16}O and ^{20}Ne + ^{12}C. Z. Phys. A: Hadrons Nucl. **346**(3), 189–200 (1993)
15. C.R. Brune, R.W. Kavanagh, C. Rolfs, ^{3}H(α, γ)^{7}Li reaction at low energies. Phys. Rev. C **50**, 2205–2218 (1994)
16. S.T. Butler, *Nuclear Stripping Reactions* (Horwitz Publications, Sydney, 1957)

© Springer Nature Switzerland AG 2019
V. Zagrebaev, *Heavy Ion Reactions at Low Energies*, Lecture Notes in Physics 963,
https://doi.org/10.1007/978-3-030-27217-3

17. L. Corradi, J.H. He, D. Ackermann, A.M. Stefanini, A. Pisent, S. Beghini, G. Montagnoli, F. Scarlassara, G.F. Segato, G. Pollarolo, C.H. Dasso, A. Winther, Multinucleon transfer reactions in ^{40}Ca + ^{124}Sn. Phys. Rev. C **54**, 201–205 (1996)
18. L. Corradi, G. Pollarolo, S. Szilner, Multinucleon transfer processes in heavy-ion reactions. J. Phys. G: Nucl. Part. Phys. **36**(11), 113101, 2009
19. K.T.R. Davies, A.J. Sierk, J.R. Nix, Effect of viscosity on the dynamics of fission. Phys. Rev. C **13**, 2385–2403 (1976)
20. A.S. Dem'yanova, A.A. Ogloblin, S.N. Ershov, F.A. Gareev, R.S. Kurmanov, E.F. Svinareva, S.A. Goncharov, V.V. Adodin, N. Burtebaev, J.M. Bang, J.S. Vaagen, Rainbows in nuclear reactions and the optical potential. Phys. Scr. **1990**(T32), 89 (1990)
21. J.F. Dicello, G. Igo, W.T. Leland, F.G. Perey, Differential elastic cross sections for protons between 10 and 22 MeV on ^{40}Ca. Phys. Rev. C **4**, 1130–1138 (1971)
22. J.M. Eisenberg, W. Greiner, *Microscopic Theory of the Nucleus* (North-Holland, Amsterdam, 1972)
23. M.A.G. Fernandes, F.E. Bertrand, R.L. Auble, R.O. Sayer, B.L. Burks, D.J. Horen, E.E. Gross, J.L. Blankenship, D. Shapira, M. Beckerman, Single-nucleon transfer reactions induced by 376 MeV ^{17}O on ^{208}Pb. Phys. Rev. C **36**, 108–114 (1987)
24. H. Feshbach, Unified theory of nuclear reactions. Ann. Phys. **5**(4), 357–390 (1958)
25. H. Feshbach, A unified theory of nuclear reactions. Ann. Phys. **19**, 287–313 (1962)
26. H. Feshbach, C.E. Porter, V.F. Weisskopf, Model for nuclear reactions with neutrons. Phys. Rev. **96**, 448–464 (1954)
27. A. Gobbi et al., in *Proceedings of the International School of Physics "Enrico Fermi" (Varenna, 1979)* (North-Holland, Amsterdam, 1979), p. 1
28. K. Hagino, N. Rowley, A.T. Kruppa, A program for coupled-channel calculations with all order couplings for heavy-ion fusion reactions. Comput. Phys. Commun. **123**, 143–152 (1999)
29. B. Hahn, D.G. Ravenhall, Robert Hofstadter, High-energy electron scattering and the charge distributions of selected nuclei. Phys. Rev. **101**, 1131–1142 (1956)
30. D.L. Hill, J.A. Wheeler, Nuclear constitution and the interpretation of fission phenomena. Phys. Rev. **89**, 1102–1145 (1953)
31. D.L. Hillis, E.E. Gross, D.C. Hensley, C.R. Bingham, F.T. Baker, A. Scott, Multistep effects in the elastic and inelastic scattering of 70.4-MeV ^{12}C ions from the even neodymium isotopes. Phys. Rev. C **16**, 1467–1482 (1977)
32. S. Hofmann et al., *GSI Report*, vol. 2012-1 (GSI Helmholtzzentrum fur Schwerionenforschung, Darmstadt, 2012), p. 205
33. S. Hofmann, G. Münzenberg, The discovery of the heaviest elements. Rev. Mod. Phys. **72**, 733–767 (2000)
34. A.V. Ignatyuk, *Statistical Properties of Excited Atomic Nuclei* (Ehnergoatomizdat, USSR, 1983)
35. M. Itkis et al., in *Proceedings on Fusion Dynamics at the Extremes* (World Scientific, Singapore, 2001)
36. M.G. Itkis, J. Aysto, S. Beghini, A.A. Bogachev, L. Corradi, O. Dorvaux, A. Gadea, G. Giardina, F. Hanappe, I.M. Itkis, M. Jandel, J. Kliman, S.V. Khlebnikov, G.N. Kniajeva, N.A. Kondratiev, E.M. Kozulin, L. Krupa, A. Latina, T. Materna, G. Montagnoli, Yu.Ts. Oganessian, I.V. Pokrovsky, E.V. Prokhorova, N. Rowley, V.A. Rubchenya, A.Ya. Rusanov, R.N. Sagaidak, F. Scarlassara, A.M. Stefanini, L. Stuttge, S. Szilner, M. Trotta, W.H. Trzaska, D.N. Vakhtin, A.M. Vinodkumar, V.M. Voskressenski, V.I. Zagrebaev, Shell effects in fission and quasi-fission of heavy and superheavy nuclei. Nucl. Phys. A **734**, 136–147 (2004)
37. A.R. Junghans, M. de Jong, H.-G. Clerc, A.V. Ignatyuk, G.A. Kudyaev, K.-H. Schmidt, Projectile-fragment yields as a probe for the collective enhancement in the nuclear level density. Nucl. Phys. A **629**(3), 635–655 (1998)
38. Y.M. Kazarinov, Y.N. Simonov, Scattering of 200 MeV neutrons by protons. Sov. Phys. JETP **16**, 24–26 (1963)
39. J. Khuyagbaatar, A. Yakushev, C.E. Dullmann, D. Ackermann, L.-L. Andersson, M. Asai, M. Block, R.A. Boll, H. Brand, D.M. Cox, M. Dasgupta, X. Derkx, A.D. Nitto, K. Eberhardt,

J. Even, M. Evers, C. Fahlander, U. Forsberg, J.M. Gates, N. Gharibyan, P. Golubev, K.E. Gregorich, J.H. Hamilton, W. Hartmann, R.-D. Herzberg, F. Heβberger, D.J. Hinde, J. Hoffmann, R. Hollinger, A. Hubner, E. Jager, B. Kindler, J.V. Kratz, J. Krier, N. Kurz, M. Laatiaoui, S. Lahiri, R. Lang, B. Lommel, M. Maiti, K. Miernik, S. Minami, A. Mistry, C. Mokry, H. Nitsche, J.P. Omtvedt, G.K. Pang, P. Papadakis, D. Renisch, J. Roberto, D. Rudolph, J. Runke, K. Rykaczewski, L.G. Sarmiento, M. Schadel, B. Schausten, A. Semchenkov, D.A. Shaughnessy, P. Steinegger, J. Steiner, E.E. Tereshatov, P. Thorle-Pospiech, K. Tinschert, T. Torres De Heidenreich, N. Trautmann, A. Turler, J. Uusitalo, D.E. Ward, M. Wegrzecki, N. Wiehl, S.M. Van Cleve, V.Yakusheva, *Study of the* $^{48}Ca + ^{249}Bk$ *fusion reaction leading to element* $Z = 117$: *long-lived* α-*decaying* ^{270}Db *and discovery of* ^{266}Lr, vol. 2014-1 of *GSI Report* (GSI Helmholtzzentrum fur Schwerionenforschung, Darmstadt, 2014), p. 125

40. P. Kunz, DWUCK5: Distorted wave born approximation – DWUCK5 computer code (1990). http://www.oecd-nea.org/tools/abstract/detail/nesc9872/

41. W.A.S. Lamb, R.E. Hester, Radiative capture of protons in carbon from 80 to 126 keV. Phys. Rev. **107**, 550–553 (1957)

42. J.R. Leigh, M. Dasgupta, D.J. Hinde, J.C. Mein, C.R. Morton, R.C. Lemmon, J.P. Lestone, J.O. Newton, H. Timmers, J.X. Wei, N. Rowley, Barrier distributions from the fusion of oxygen ions with ^{144,148,154}Sm and ^{186}W. Phys. Rev. C **52**, 3151–3166 (1995)

43. J. Maruhn, W. Greiner, The asymmetrie two center shell model. Z. Angew. Phys. **251**(5), 431–457 (1972)

44. P. Moller, J.R. Nix, W.D. Myers, W.J. Swiatecki, Nuclear ground-state masses and deformations. At. Data Nucl. Data Tables **59**(2), 185–381 (1995)

45. L.G. Moretto, J.S. Sventek, A theoretical approach to the problem of partial equilibration in heavy ion reactions. Phys. Lett. B **58**(1), 26–30 (1975)

46. K. Morita, K. Morimoto, D. Kaji, T. Akiyama, S.-I. Goto, H. Haba, E. Ideguchi, K. Katori, H. Koura, H. Kikunaga, H. Kudo, T. Ohnishi, A. Ozawa, N. Sato, T. Suda, K. Sueki, F. Tokanai, T. Yamaguchi, A. Yoneda, A. Yoshida, Observation of second decay chain from 278113. J. Phys. Soc. Jpn. **76**(4), 045001 (2007)

47. C.R. Morton, D.J. Hinde, J.R. Leigh, J.P. Lestone, M. Dasgupta, J.C. Mein, J.O. Newton, H. Timmers, Resolution of the anomalous fission fragment anisotropies for the ^{16}O+^{208}pb reaction. Phys. Rev. C **52**, 243–251 (1995)

48. T. Murakami, C.-C. Sahm, R. Vandenbosch, D.D. Leach, A. Ray, M.J. Murphy, Fission probes of sub-barrier fusion cross section enhancements and spin distribution broadening. Phys. Rev. C **34**, 1353–1365 (1986)

49. W.D. Myers, W.J. Swiatecki, Anomalies in nuclear masses. Ark. Phys. **36**, 343 (1967)

50. R.S. Naik, W. Loveland, P.H. Sprunger, A.M. Vinodkumar, D. Peterson, C.L. Jiang, S. Zhu, X. Tang, E.F. Moore, P. Chowdhury, Measurement of the fusion probability P_{cn} for the reaction of ^{50}Ti with ^{208}Pb. Phys. Rev. C **76**, 054604 (2007)

51. R.G. Newton, *Scattering Theory of Waves and Particles* (Springer, New York, 1982)

52. Y. Oganessian, Heaviest nuclei from 48Ca-induced reactions. J. Phys. G: Nucl. Part. Phys. **34**(4), R165 (2007)

53. Y.E. Penionzhkevich, V.I. Zagrebaev, S.M. Lukyanov, R. Kalpakchieva, Deep sub-barrier fusion enhancement in the ^{6}He $+^{206}$ Pb reaction. Phys. Rev. Lett. **96**, 162701 (2006)

54. K.E. Rehm, A.M. van den Berg, J.J. Kolata, D.G. Kovar, W. Kutschera, G. Rosner, G.S.F. Stephans, J. L. Yntema, Transition from quasi-elastic to deep-inelastic reactions in the ^{48}Ti + ^{208}Pb system. Phys. Rev. C **37**, 2629–2646 (1988)

55. C. Rolfs, R.E. Azuma, Interference effects in ^{12}C(p,γ)^{13}Nand direct capture to unbound states. Nucl. Phys. A **227**(2), 291–308 (1974)

56. N. Rowley, G.R. Satchler, P.H. Stelson, On the distribution of barriers interpretation of heavy-ion fusion. Phys. Lett. B **254**(1), 25–29 (1991)

57. R.N. Sagaidak, G.N. Kniajeva, I.M. Itkis, M.G. Itkis, N.A. Kondratiev, E.M. Kozulin, I.V. Pokrovsky, A.I. Svirikhin, V.M. Voskressensky, A.V. Yeremin, L. Corradi, A. Gadea, A. Latina, A.M. Stefanini, S. Szilner, M. Trotta, A.M. Vinodkumar, S. Beghini, G. Montagnoli, F. Scarlassara, D. Ackermann, F. Hanappe, N. Rowley, L. Stuttgé, Fusion suppression in mass-asymmetric reactions leading to Ra compound nuclei. Phys. Rev. C **68**, 014603 (2003)

58. G. Satchler, W. Love, Folding model potentials from realistic interactions for heavy-ion scattering. Phys. Rep. **55**, 183–254 (1979)

59. M. Schädel, J.V. Kratz, H. Ahrens, W. Brüchle, G. Franz, H. Gäggeler, I. Warnecke, G. Wirth, G. Herrmann, N. Trautmann, M. Weis, Isotope distributions in the reaction of ^{238}U with ^{238}U. Phys. Rev. Lett. **41**, 469–472 (1978)

60. W.Q. Shen, J. Albinski, A. Gobbi, S. Gralla, K.D. Hildenbrand, N. Herrmann, J. Kuzminski, W.F.J. Müller, H. Stelzer, J. Tke, B.B. Back, S. Bjrnholm, S.P. Srensen, Fission and quasifission in U-induced reactions. Phys. Rev. C **36**, 115–142 (1987)

61. T. Sikkeland, J. Maly, D.F. Lebeck, Evaporation of 3 to 8 neutrons in reactions between ^{12}C and various uranium nuclides. Phys. Rev. **169**, 1000–1006 (1968)

62. K.A. Snover, Giant resonances in excited nuclei. Annu. Rev. Nucl. Part. Sci. **36**(1), 545–603 (1986)

63. H. Sohlbach, H. Freiesleben, W.F.W. Schneider, D. Schüll, B. Kohlmeyer, M. Marinescu, F. Pühlhofer, Inelastic excitation and nucleon transfer in quasielastic reactions between ^{86}Kr and ^{208}Pb at 18.2 MeV/u beam energy. Z. Phys. A: At. Nucl. **328**(2), 205–217 (1987)

64. A.M. Stefanini, L. Corradi, A.M. Vinodkumar, Y. Feng, F. Scarlassara, G. Montagnoli, S. Beghini, M. Bisogno, Near-barrier fusion of ^{36}S+^{90,96}Zr : the effect of the strong octupole vibration of ^{96}Zr. Phys. Rev. C **62**, 014601 (2000)

65. V.M. Strutinsky, Shells in deformed nuclei. Nucl. Phys. A **122**(1), 1–33 (1968)

66. G.M. Ter-Akopian, A.M. Rodin, A.S. Fomichev, S.I. Sidorchuk, S.V. Stepantsov, R. Wolski, M.L. Chelnokov, V.A. Gorshkov, A.Yu. Lavrentev, V.I. Zagrebaev, Yu.Ts. Oganessian. Two-neutron exchange observed in the ^{6}He + ^{4}He reaction. Search for the di-neutron configuration of ^{6}He. Phys. Lett. B **426**(3), 251–256 (1998)

67. I.J. Thompson, Coupled reaction channels calculations in nuclear physics. Comput. Phys. Rep. **7**(4), 167–212 (1988)

68. R. Vandenbosch, J. R. Huizenga, *Nuclear Fission* (Academic Press, New York, 1973)

69. W. von Oertzen, H.G. Bohlen, B. Gebauer, R. Künkel, F. Pühlhofer, D. Scühll, Quasi-elastic neutron transfer and pairing effects in the interaction of heavy nuclei. Z. Phys. A: At. Nucl. **326**(4), 463–481 (1987)

70. C.F.v. Weizsäcker, Zur theorie der kernmassen. Z. Angew. Phys. **96**(7), 431–458 (1935)

71. J. Wilczynski, Nuclear molecules and nuclear friction. Phys. Lett. B **47**(6), 484–486 (1973)

72. A. Winther, Grazing reactions in collisions between heavy nuclei. Nucl. Phys. A **572**(1), 191–235 (1994)

73. A. Winther, Dissipation, polarization and fluctuation in grazing heavy-ion collisions and the boundary to the chaotic regime. Nucl. Phys. A **594**(2), 203–245 (1995)

74. A. Winther, A fortran program for estimating reactions in collision between heavy nuclei. Technical report, The Niels Bohr Institutet (2002). http://personalpages.to.infn.it/~nanni/grazing/

75. H.J. Wollersheim, W.W. Wilcke, J.R. Birkelund, J.R. Huizenga, W.U. Schröder, H. Freiesleben, D. Hilscher, ^{209}Bi + ^{136}Xe reaction at E_{lab} = 1422 MeV. Phys. Rev. C **24**, 2114–2126 (1981)

76. V.I. Zagrebaev, Semiclassical theory of direct and deep inelastic heavy ion collisions. Ann. Phys. **197**(1), 33–93 (1990)

77. A.V. Karpov, A.S. Denikin, M.A. Naumenko, A.P. Alekseev, V.A. Rachkov, V.V. Samarin, V.V. Saiko, V.I. Zagrebaev, NRV web knowledge base on low-energy nuclear physics. Nucl. Instrum. Methods Phys. Res. Sect. A **859**, 112–124 (2017)

78. V.I. Zagrebaev, A.V. Karpov, Y. Aritomo, M.A. Naumenko, W. Greiner, Potential energy of heavy nuclear system in fusion-fission processes. Phys. Elementary Particles At. Nucl. **38**, 893–938 (2007)

Printed in the United States
By Bookmasters